KU-735-572

BIOCHEMISTRY OF ANTIMICROBIAL ACTION

T. J. Franklin and G. A. Snow

in collaboration with

K. J. Barrett-Bee and R. D. Nolan

ICI Pharmaceuticals Division
Alderley Park
Macclesfield
Cheshire

FOURTH EDITION

London   New York
CHAPMAN AND HALL

First published in 1971 by Chapman and Hall Ltd
11 New Fetter Lane, London EC4P 4EE
Second edition 1975, third edition 1981
Reprinted 1985,
Fourth edition 1989
Published in the USA by Chapman and Hall
29 West 35th Street, New York NY 10001
© 1971, 1975, 1981, 1989 T. J. Franklin and G. A. Snow

Typeset in 10/12pt Baskerville by Acorn Bookwork,
Salisbury, Wiltshire
Printed in Great Britain by St Edmundsbury Press,
Bury St Edmunds, Suffolk

ISBN 0 412 30250 0 (Hb)
ISBN 0 412 30260 8 (Pb)

This title is available in both hardbound and paperback editions. The paperback
edition is sold subject to the condition that it shall not, by way of trade or otherwise,
be lent, resold, hired out, or otherwise circulated without the publisher's prior
consent in any form of binding or cover other than that in which it is published and
without a similar condition including this condition being imposed on the subsequent
purchaser.
All rights reserved. No part of this book may be reprinted, or reproduced or utilized in
any form or by any electronic, mechanical or other means, now known or hereafter
invented, including photocopying and recording, or in any information storage and
retrieval system, without permission in writing from the publisher.

British Library Cataloguing in Publication Data

Franklin, T. J. (Trevor John)
    Biochemistry of antimicrobial action
    4th ed.
    1. Antimicrobials. Biochemistry
    I. Title   II. Snow, G. A. (George Alan)
    615'.7

    ISBN 0-412-30250-0
    ISBN 0-412-30260-8 Pbk

Library of Congress Cataloging in Publication Data

Franklin, T. J.
    Biochemistry of antimicrobial action / T. J. Franklin and G. A. Snow; in
collaboration with K. J. Barrett-Bee and R. D. Nolan.—4th ed.   p.   cm.
    Includes bibliographies and index.
    ISBN 0-412-30250-0.   ISBN 0-412-30260-8 (pbk.)
    1. Anti-infective agents—Physiological effect.   2. Microbial
metabolism.   3. Micro-organisms—Effect of antibiotics on.   I. Snow,
G. A.   II. Title.
    [DNLM: 1. Antibiotics—pharmacology.   2. Anti-Infective agents—
pharmacology.   QV 350 F834b]
    QR97.A57F73 1989
    616.9'0461—dc19
    DNLM/DLC
    for Library of Congress   88-20222
                                                                          CIP

Contents

Preface

The rapid advances made in the study of the synthesis, structure and function of biological macromolecules in the last fifteen years have enabled scientists concerned with antimicrobial agents to achieve a considerable measure of understanding of how these substances inhibit cell growth and division. The use of antimicrobial agents as highly specific inhibitors has in turn substantially assisted the investigation of complex biochemical processes. The literature in this field is so extensive however, that we considered an attempt should be made to draw together in an introductory book the more significant studies of recent years. This book, which is in fact based on lecture courses given by us to undergraduates at Liverpool and Manchester Universities, is therefore intended as an introduction to the biochemistry of antimicrobial action for advanced students in many disciplines. We hope that it may also be useful to established scientists who are new to this area of research.

The book is concerned with a discussion of medically important antimicrobial compounds and also a number of agents that, although having no medical uses, have proved invaluable as research tools in biochemistry. Our aim has been to present the available information in a simple and readable way, emphasizing the established facts rather than more controversial material. Whenever possible, however, we have indicated the gaps in the present knowledge of the subject where further information is required. We have avoided the use of literature references in the text; instead we have included short lists of key articles and books for further reading at the end of each chapter.

We have drawn on the work of many scientists and we are especially pleased to express our thanks to those who have given us permission to reproduce their original diagrams and photographs. We are also grateful to the Pharmaceuticals Division of Imperial Chemicals Industries Ltd, for providing the necessary facilities for the preparation of this book.

Abbreviations used without definition for common biochemical substances are those recommended by the Biochemical Journal (1970).

T. J. FRANKLIN
G. A. SNOW

June 1970

vii

Preface to the fourth edition

The decision to prepare a fourth edition was prompted by developments in a number of areas. Interest in the β-lactam drugs has been sustained by further ingenious exploitation of the potential of this group of drugs to provide evermore powerful treatments of bacterial infections. The bacteria themselves, however, have proved resourceful in deploying new modes of resistance to recently devised β-lactams. There has been a resurgence of interest in quinolone antibacterials in recent years and closer attention is being paid to the chemotherapy of fungal and viral infections, prompted largely by the AIDS epidemic. Knowledge of the permeability barriers of the Gram-negative outer membrane and of the biology of drug-resistance plasmids has expanded dramatically. We have dealt with these and many other new developments within the limited space available to us. In the interests of logical presentation we have rearranged the distribution of some material in Chapters 2 and 6. Over the years we have received many helpful comments and criticisms which have influenced subsequent editions. We hope that our readers will continue to let us know their views.

December 1987

T. J. FRANKLIN
G. A. SNOW
K. J. BARRETT-BEE
R. D. NOLAN

1 The development of antimicrobial agents, past, present and future

THE SOCIAL AND ECONOMIC IMPORTANCE OF ANTIMICROBIAL AGENTS

Few developments in the history of medicine have had such a profound effect upon human life and society as the development of the power to control infections due to micro-organisms. The impact has been felt in many ways. The developed countries have been freed from the devastating plagues that in earlier days decimated the population leaving a trail of misery and social disruption. No longer is surgery a desperate gamble with human life. The perils of childbirth are greatly lessened with the control of puerperal fever. The death of children and young adults from meningitis, tuberculosis and septicaemia, once commonplace, is now unusual. All these benefits are now accepted without question, but they have changed the fabric of our lives. The control of infectious disease has also brought its own social and medical problems. Within our own highly developed society the conquest of pneumonia, sometimes called 'the old man's friend', has helped to raise the expectation of life. People live longer and the aged form an increasing part of our population. Even so microbial pathogens still have the capacity to spring unpleasant surprises on the world. The current epidemic of the viral infection that causes the Acquired Immune Deficiency Syndrome, AIDS, will require the combined efforts of public health officials, educationists, research scientists and the medical profession to contain it.

The ability to control infectious disease has emphasized our lesser success in dealing with degenerative diseases, and an extended life span is a dubious advantage if it is accompanied by failing faculties and the feeling of being a burden on an uncaring society. In the poorer and less-developed countries the control of infection has begun to show its effects. Nearly all the diseases affecting many millions of people, such as tuberculosis, malaria, leprosy and schistosomiasis could in principle be abolished or reduced to minor proportions. The major obstacle to achieving this is the cost. The continued

1

prevalence of these diseases is a reflection on the disparity still remaining between the 'haves' and the 'have nots'. Even so, considerable progress has been made. For example, the terrible scourge of smallpox has been eradicated in the past ten years through a determined and comprehensive programme of vaccination against the disease. An obvious consequence of success in controlling infections in the 'Third World' is a sudden increase in populations. Communities where endemic disease and periodic plagues previously kept the death rate in balance with the birth rate are now expanding at an alarming rate. This effect may well increase as measures against infection continue to improve. Population control is essential if these communities are not to suffer the effects of gross overcrowding and undernourishment in place of their previous situation of high maternal and infantile mortality and short life expectation due to infectious disease.

In the fight against infectious disease several factors have combined to produce the remarkable achievements already outlined. The first advances were mainly the result of improved sanitation and housing. These removed some of the worst foci of infectious disease and limited the spread of infection through vermin and insect parasites or by contaminated water and food. The earliest effective direct control of infectious diseases was achieved through vaccination and similar immunological methods. These still play a large part in the control of infection today, and they are the mainstay of our attack on virus diseases which largely resist treatment by chemotherapeutic agents. The use of antimicrobial drugs for the control of infection is almost entirely a development of this century, and the most dramatic developments have taken place only since the late 1930s. The practical importance of antimicrobial agents is enormous. In the pharmaceutical industry this is probably the largest class of drugs with world sales around £7000m per annum (1986). Rarely has so complete a success been achieved in the medical field in so short a time. Before describing the biochemistry underlying the action of antimicrobial drugs we shall consider the work that made these drugs available to us and the thought that lay behind their discovery.

AN OUTLINE OF THE HISTORICAL DEVELOPMENT OF ANTIMICROBIAL AGENTS AND OF CHEMOTHERAPEUTIC THEORIES

Early remedies

Among the many traditional and folk remedies two sources of antimicrobial compounds have survived to the present day. These are cinchona bark for the treatment of malaria and ipecacuanha root for amoebic dysentery. Cinchona bark was used by the Indians of Peru for treating malaria and was introduced into European medicine by the Spaniards in the early seventeenth century. The active principle, quinine, was isolated in 1820. Quinine remained the only treatment for malaria until well into the twentieth century and still has a place in chemotherapy. Ipecacuanha root was known in Brazil and probably in Asia for its curative action in diarrhoeas and dysentery. Emetine was isolated as the active constituent in 1817 and was shown in 1891 to have a specific action against amoebic dysentery. It is still used for treating this disease. These early remedies were used along with many ineffectual nostrums without any understanding of the nature of the diseases. Malaria, for example, was thought to be caused by miasmas arising from marshy places; the significance of the blood-borne parasite was not recognized until 1883 and only in 1899 was the anophelene mosquito proved to be the specific insect vector.

Antiseptics and disinfectants

The use of disinfectants and antiseptics also preceded understanding of their action, and seems to have arisen from the observation that certain substances stopped putrefaction of meat or rotting of wood. The term 'antiseptic' itself was apparently first used by Pringle in 1750 to describe substances that prevent putrefaction. The idea was eventually applied to the treatment of suppurating wounds. Mercuric chloride was used by Arabian physicians in the Middle Ages for preventing sepsis in open wounds. However, it was not until the nineteenth century that antiseptics came into general use in medicine. Chlorinated soda, essentially hypochlorite, was introduced in 1825 by Labarraque for the treatment of infected wounds, and tincture of iodine was first used in 1839. One of the earliest examples of disinfection used in preventing the spread of infectious disease was recorded by Oliver Wendel Holmes in 1835. He regularly washed his hands in a solution of chloride of lime when dealing with cases of puerperal fever and thereby greatly reduced the incidence of fresh infections as did Ignaz Semmelweiss in Vienna a few years later. These pioneer attempts at antisepsis were not generally accepted until Pasteur's publication in 1863 of the microbial origin of putrefaction. This led to an understanding of the origin of infection and suggested the rationale for its prevention. As so often in the history of medicine a change of practice depended upon the personality and persistence of one man. In antiseptics this man was Lister. He chose phenol, the antiseptic which had been introduced by Lemaire in 1860, and applied it vigorously in surgery. A 2.5% solution was used for dressing wounds and twice this concentration for sterilizing instruments. Later he used a spray of phenol solution to produce an essentially sterile environment for carrying out surgical operations. The previous state of surgery had been deplorable; wounds usually became infected and the mortality rate was appalling. The effect of Lister's measures was revolutionary and the antiseptic technique opened the way to great surgical advances. Even at this time, about 1870, the use of antiseptics was still empirical. An understanding of their function began with the work of Koch who from 1881 onwards introduced the techniques on which modern bacteriology has been built. He perfected methods of obtaining pure cultures of bacteria and of growing them on solid media and he demonstrated practical methods of sterile working. Once it became possible to handle bacteria in a controlled environment the action of disinfectants and antiseptics could be studied. The pioneer work on the scientific approach to this subject was published by Kronig and Paul in 1897.

Since that time the history of antiseptics has been of steady but unspectacular improvement. Many of the traditional antiseptics have continued in use in refined forms. The phenols have been modified and made more acceptable for general use. Acriflavine, introduced in 1913, was the first of a number of basic antiseptics. It had many years of use but has been largely displaced in the last three decades by colourless cationic antiseptics. In surgery the antiseptic era gave place to the aseptic era in which the emphasis is on the avoidance of bacterial contamination rather than on killing bacteria already present. All the same, infection of surgical wounds is a constant risk and antiseptics are still used as an extra precaution or second line of defence. Surgical staff also 'scrub up' with mild antiseptic solutions before entering the operating theatre. Disinfectants play an important part in the hygiene of milking sheds, broiler houses and other places where strict asepsis is impracticable.

The beginnings of chemotherapy

The publications of Pasteur and Koch firmly established that microorganisms are the cause of infectious disease, though for some diseases the causative organism still remained to be discovered. It was also known that bacteria are killed by various antiseptics and disinfectants. Not surprisingly attempts were made to kill micro-organisms within the body and so to end an infection. Koch himself carried out some experiments with this aim. He had shown the action of mercuric chloride against the anthrax bacillus; it is one of the few disinfectants that kill the particularly tough spores of this organism. He therefore tried to cure animals of anthrax infection by injecting mercuric chloride. Unfortunately the animals died of mercury poisoning and their organs still contained infectious anthrax bacilli. A slightly more successful attempt to cure an infection with a toxic agent was made by Lindgard in 1893. He treated horses suffering from surra, a disease now known to be caused by trypanosomes, with arsenious oxide. There was some improvement of the disease, but the compound was too toxic to be generally useful.

Chemotherapy however really began with Paul Ehrlich. During the ten years from 1902 onwards Ehrlich's work foreshadowed almost all the concepts which have governed subsequent work on synthetic antimicrobial agents. His first ideas arose from studies with 'vital stains': dyestuffs that were taken up selectively by living tissue. One such dye was Methylene Blue, which in the animal body is concentrated in nervous tissue. Ehrlich showed that the same dye was readily taken up by the malaria parasites in the blood so that they become deeply stained. Consequently Methylene Blue was tried against human malaria and showed some effect, though not sufficient to make it a useful treatment. Nevertheless this minor success started a line of thought that was to prove of the greatest significance. Ehrlich believed that antimicrobial agents must be essentially toxic compounds and that they must bind to the micro-organism in order to exert their action. The problem was to discover compounds having a selective action against the microbial cell compared with the cells of the host animal. Starting from Methylene Blue Ehrlich began to search for other dyestuffs that would affect protozoal diseases. In 1904, after testing hundreds of available dyes, he eventually found one that was effective against trypanosomiasis in horses. This compound called Trypan Red was a significant landmark in the treatment of microbial infections since it was the first man-made compound that produced a curative effect.

However, it was not in the field of dyestuffs that Ehrlich achieved his greatest success. Following the early work on the treatment of trypanosomiasis with arsenious oxide, Koch tested the organic arsenical atoxyl (Fig. 1.1). This compound produced the first cures of sleeping sickness, a human trypanosomal disease. The compound however produced unfortunate side

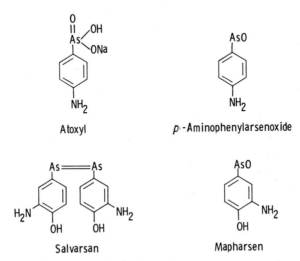

Figure 1.1 Arsenical compounds used in the treatment of trypanosomiasis or syphilis.

effects, some patients developing optic atrophy. The curative success of this compound stimulated Ehrlich to make other related arsenicals. He tested these on mice infected experimentally with trypanosomiasis, and showed that curative action did not run parallel with toxicity to the mice. This suggested that if enough compounds were made some would have sufficiently low toxicity to be safe as chemotherapeutic agents. Ehrlich continued his search for compounds active against various micro-organisms and showed that arsenicals were active against the causative organism of syphilis. He began a massive search for an organoarsenical compound that could be used in the treatment of this disease and eventually in 1910 discovered the famous drug salvarsan (Fig. 1.1). This drug and its derivative neo-salvarsan became the standard treatment for syphilis. Coupled with bismuth therapy they remained in use until supplanted by penicillin in 1945. This was the most spectacular practical achievement of Ehrlich's career, but scientifically he is remembered at least as much for his wealth of ideas that have inspired workers in the field of chemotherapy down to the present day. These ideas are so important that they deserve separate consideration.

The debt of chemotherapy to Ehrlich

The very term chemotherapy was invented by Ehrlich and expressed his belief that infectious disease could be combated by treatment with synthetic chemicals. Successes since his day have entirely justified his faith in this possibility. He postulated that cells possess chemical receptors which are concerned with the uptake of nutrients. Drugs that affect the cell must bind to one or other of these receptors. The toxicity of a drug is determined firstly by its distribution in the body. However, in the treatment of an infection it is the binding to the parasite relative to the host cell that determines the effectiveness of the compound. Thus Ehrlich recognized the importance of quantitative measurement of the relationship between the dose of a compound required to produce a therapeutic effect and the dose that will cause toxic reactions. Such measurements are still of prime importance in chemotherapy today. Ehrlich's view of cell receptors also has its counterpart in modern views of membrane transport.

Ehrlich pioneered methods that have since become the mainstay of the search for new drugs. One aspect of his approach was the use of 'screening'. This is the application of a relatively simple test to large numbers of compounds in order to obtain evidence of biological activity in types of chemical structure not previously examined. The second of Ehrlich's methods was the deliberate synthesis of chemical variants of a compound known to have the required activity. The new compounds were examined for increased activity or for improvements in some other property such as reduced toxicity. Any improvement found was used as a guide to further synthesis eventually arriving, by a series of steps, at the best possible compound. These methods are now so well accepted that their novelty in Ehrlich's day can easily be forgotten. They depend on the thesis that a useful drug possesses an ideal combination of structural features which cannot be predicted at the outset. A compound having a sufficiently close structural approximation will show some degree of activity, and can therefore act as a 'lead' towards the best attainable structure.

According to Ehrlich a chemotherapeutic substance has two functional features, the 'haptophore' or binding group which enables the compound to attach itself to the cell receptors, and the 'toxophore' or toxic group that brings about an adverse effect on the cell. This idea has had a continuing influence in subsequent years. In cancer chemotherapy it has frequently been used in attempts to bring about specific concentration of toxic agents or antimetabolites in tumour cells. In antimicrobial research it has helped to explain some features of the biochemical action of antimicrobial compounds.

Ehrlich also recognized that compounds acting on microbial infection need not necessarily kill the invading organism. It was, he suggested, sufficient to prevent substantial multiplication of the infectious agent, since the normal body defences, antibodies and phagocytes would cope with foreign organisms provided that their numbers were not overwhelming. His views on this topic were based in part on his observation that isolated spirochaetes treated with low concentrations of salvarsan remained motile and were therefore apparently still alive. Nevertheless they were unable to produce an infection when they were injected into an animal body. It is a striking fact that several of today's important antibacterial drugs are bacteriostatic rather than bactericidal.

Another feature of Ehrlich's work was his recognition of the possibility that drugs may be activated by metabolism in the body. This suggestion was prompted by the observation that the compound atoxyl was active against trypanosomal infections but was inactive against isolated trypanosomes. His explanation was that atoxyl was reduced in the body to the much more toxic p-aminophenylarsenoxide (Fig. 1.1). More recent work shows that atoxyl and other related arsenic acids are not readily reduced to arsenoxides in the body; however, local reduction by the parasite remains a possibility. Ehrlich, surprisingly, did not recognize that his own compound salvarsan would undergo metabolic cleavage. In animals it gives rise to the arsenoxide as the first of a series of metabolites. This compound was introduced into medicine in 1932 under the name mapharsen (Fig. 1.1); its toxicity is rather high, but it has sufficient selectivity to give it useful chemotherapeutic properties. Other examples of activation through metabolism have been discovered in more recent times.

Ehrlich also drew attention to the problem of resistance of micro-organisms towards chemotherapeutic compounds. He noticed it in the treatment of trypanosomes with parafuchsin and later with Trypan Red and atoxyl. He found that resistance extended to other compounds chemically related to the original three, but there was no cross-resistance between the groups. In Ehrlich's view this was evidence that each of these compounds was affecting a separate receptor. Independent resistance to different drugs later became a commonplace in antimicrobial therapy. Ehrlich's view of the nature of resistance is also interesting. He found that trypanosomes resistant to Trypan Red absorbed less of the dyestuff than sensitive strains, and he postulated that the receptors in resistant organisms had a diminished affinity for the dye. This mechanism corresponds to one of the currently accepted types of resistance in micro-organisms (Chapter 8).

Several useful antimicrobial drugs arose in later years as an extension of Ehrlich's work. The most notable are suramin, a development from Trypan Red, and mepacrine (also known as quinacrine or atebrin) indirectly from Methylene Blue (Fig. 1.2). Suramin, introduced in 1920, is a colourless compound having a useful action against human trypanosomiasis. Its particular value lay in its relative safety compared with other antimicrobial drugs of the period. It was the first useful antimicrobial drug without a toxic metal atom and the ratio of the dose required to produce toxic symptoms to that needed for a curative effect is vastly higher than with any of the arsenicals. It is also remarkably persistent, a single dose giving protection for more than a month. Mepacrine, first marketed in 1933, was an antimalarial agent of immense value in the Second World War. It has been supplanted by other compounds partly because it causes a yellow discoloration of the skin.

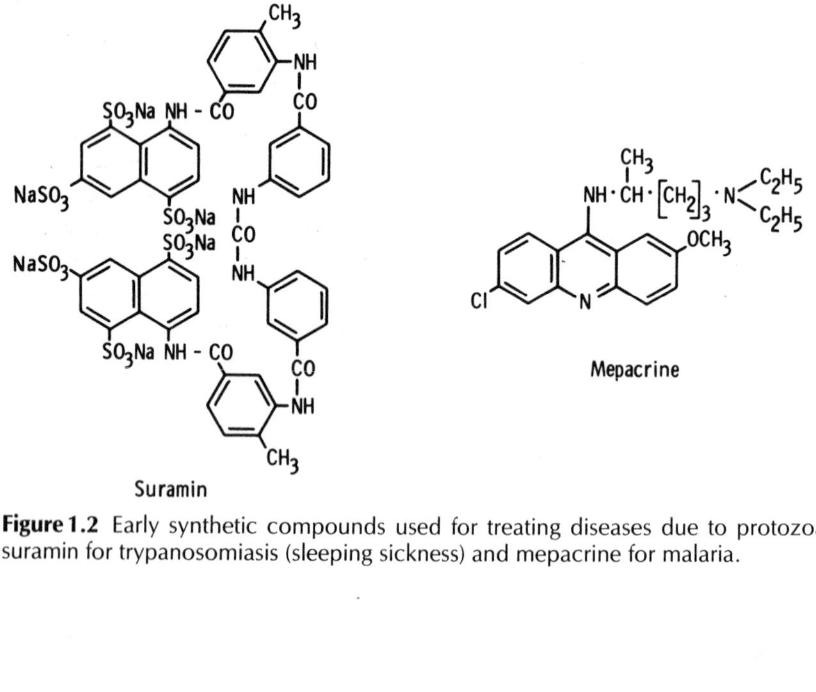

Figure 1.2 Early synthetic compounds used for treating diseases due to protozoa: suramin for trypanosomiasis (sleeping sickness) and mepacrine for malaria.

Besides these obvious descendants from Ehrlich's work the whole field of drug therapy is permeated by his ideas and many other important compounds can be traced directly or indirectly to the influence of his thought.

The treatment of bacterial infections by synthetic compounds

In spite of the successes achieved in the treatment of protozoal diseases and the spirochaetal disease syphilis the therapy of bacterial infections remained for many years an elusive and apparently unattainable goal. Ehrlich himself, in collaboration with Bechtold, made a series of phenols which showed much higher antibacterial potency than the simple phenols originally used as disinfectants. These compounds, however, had no effect on bacterial infections in animals. Other attempts were equally unsuccessful and no practical progress was made until 1935, when Domagk reported the activity of Prontosil rubrum against infections in animals. The discovery occurred in the course of a widespread research programme on the therapeutic use of dyestuffs, apparently inspired by Ehrlich's ideas. Trefouel showed Prontosil rubrum to be broken down in the body giving sulphanilamide which was in fact the effective antibacterial agent. The sulphonamides were outstandingly successful drugs. They might have been developed and used even more widely if penicillin and other antibiotics had not followed on so speedily. Surprisingly in the field of synthetic antibacterial agents few other successes have been achieved against the common bacterial infections. In part this may be due to the diminished incentive to search for new products when the antibiotics have proved to give such excellent results. However, synthetic compounds with a useful action against bacterial infections have certainly been extremely difficult to find. The only widely used synthetic compounds apart from the sulphonamides are the quinolones (Chapter 4), and trimethoprim (Chapter 4) although it should be remembered that modern β-lactam drugs (Chapter 2) are semi-synthetic variants of naturally occurring molecules. In contrast with most common bacterial infections, those caused by the mycobacteria are usually more effectively treated with synthetic agents than with antibiotics. For many years after treatment was available for streptococcal and staphylococcal infections, tuberculosis and leprosy remained untreatable by chemotherapy. The first success came with the antibiotic streptomycin, which remains part of the standard treatment for tuberculosis. Very soon afterwards however a number of chemically unrelated synthetic agents were found to be effective against this disease. The best of these are isonicotinic hydrazide and p-aminosalicylic acid, though several other types remain as second-line drugs. Apart from streptomycin and its relatives the antibiotics have played relatively little part in the treatment of tuberculosis. In leprosy the drug regularly used is a synthetic compound, 4,4′-diaminodiphenylsulphone.

The antibiotic revolution

Ever since bacteria have been cultivated on solid media contaminant organisms have occasionally appeared on the plates. Sometimes this foreign colony would be surrounded by an area in which bacterial growth was suppressed. Usually this was regarded as a mere technical nuisance, but Fleming, observing such an effect with a mould *Penicillium notatum* on a plate seeded with staphylococci, was struck by its potential importance. He showed that the mould produced a freely diffusible substance highly active against Gram-positive bacteria, and apparently of low toxicity to animals. He named it penicillin. It was however unstable and early attempts to extract it failed, so Fleming's observation lay neglected until 1939. By then the success of the sulphonamides had stimulated a renewed interest in the chemotherapy of bacterial infections. The search for other antibacterial agents now seemed a promising and exciting project and Florey and Chain selected Fleming's penicillin for re-examination. They succeeded in isolating an impure but highly active solid preparation and published their results in 1940. Evidence of its great clinical usefulness in man followed in 1941. It was now apparent that a compound of revolutionary importance in medicine had been discovered. To make it generally available for medical use, however, presented formidable problems both in research and in large scale production, especially under conditions of war-time stringency. Eventually perhaps the biggest chemical and biological joint research programme ever mounted was undertaken, involving 39 laboratories in Britain and the United States. It was an untidy operation with much duplication and overlapping of work, but it culminated in the isolation of pure penicillin, the determination of its structure and the establishment of methods for its production on a large scale. The obstacles overcome in this research were enormous. They arose mainly from the very low concentrations of penicillin in the original mould cultures and from the marked chemical instability of the product. In the course of this work the concentration of penicillin in mould culture fluids was increased 1000-fold by the isolation of improved variants of *Penicillum notatum* using selection and mutation methods and by improved conditions of culture. This tremendous improvement in yield was decisive in making large scale production practicable and ultimately cheap.

The success of penicillin quickly diverted a great deal of scientific effort towards the search for other antibiotics. The most prominent name in this development was that of Waksman who began an intensive search for antibiotics in moulds isolated from soil samples obtained from all parts of the world. Waksman's first success was streptomycin and many other antibiotics followed. Waksman's screening methods were copied in many other laboratories. Organisms of all kinds were examined and hundreds of thousands of cultures were tested. Further successes came quickly. Out of all this research some 3000 named antibiotics have been listed. Most of them, however, have defects that prevent their development as drugs. Perhaps 50 have had some sort of clinical use and very few of these are regularly employed in the therapy of infectious disease. However, among this select group and their semi-synthetic variants are compounds of such excellent qualities that treatment is now available for almost all the bacterial infections known to occur in man.

New antibiotics continue to be isolated from nature but mostly these are variants of known types, and, with some notable exceptions such as thienamycin (Chapter 2) are mostly without sufficient advantage to favour their use in place of well established compounds.

The most significant development of recent years has been the use of chemical modification to extend and improve the action of naturally occurring antibiotics. This was first done successfully with the penicillins (Chapter 2) and has since been extended to several other classes of antibiotics. The current emphasis in the development of antibiotics has swung away from the search for novel types. Instead research concentrates on taking antibiotics of known excellence and tailoring them to produce compounds with some special advantage, such as immunity to attack by bacterial enzymes and improved antibacterial spectrum.

The synthetic antibacterials and the antibiotics illustrate a phenomenon that is not uncommon in the development of science. Sulphanilamide, isonicotinic hydrazide and p-aminosalicylic acid were all compounds known to chemistry long before their antibacterial potency was realized. Many of the antibiotics are produced by commonly occurring micro-organisms and are relatively easy to isolate. There is no technical reason why they should not have been found many years earlier. It required the stimulus of the chance discovery of the activity of the sulphonamides to encourage the large scale search that in fifteen years transformed the medical treatment of bacterial infection.

Antifungal and antiviral drugs

The diversity of fungal pathogens which attack man and his domesticated animals is considerably smaller than that of bacteria. Nevertheless fungi cause infections ranging from the trivial and inconvenient to those causing major illness and death. Fungal infections have assumed greater importance in recent years because of the increased number of medical conditions in which host immunity is compromised. Cytotoxic drugs used to combat cancer, organ graft rejection and AIDS damage the immune system and increase susceptibility to infection by opportunist pathogens that rarely cause illness in healthy individuals. Fungi as eukaryotes have much more biochemistry in common with mammalian cells than bacteria and therefore pose a serious challenge to chemotherapy. Specificity of action is more difficult to achieve. Few antibiotics are useful against fungal infections and attention has concentrated on devising synthetic agents. Advantage has been taken of progress in devising compounds for the treatment of fungal infections of plants to produce reasonably safe and effective drugs of the azole type (Chapter 6) for human fungal infections.

Enormous strides have been made in the control of viral infections through the use of vaccines. As mentioned previously, smallpox appears to have been eradicated throughout the world. In the developed countries at least the seasonal epidemics of poliomyelitis that were the cause of so much fear and suffering thirty years ago have largely disappeared. But despite these and other vaccine-led successes against viral infections, not all such infections can be so effectively controlled by mass vaccination programmes. The bewildering diversity of common cold viruses, the ever-shifting antigenic profiles of influenza viruses and the insidious nature of the viruses that cause AIDS are just three examples of diseases that will not yield readily to the vaccine approach. Attention is therefore being focused on finding drugs that specifically arrest viral replication, a formidable challenge since viruses partially parasitize the biochemistry of the host cells. The encouraging progress that is being made is outlined in Chapter 6.

REASONS FOR STUDYING THE BIOCHEMISTRY OF ANTIMICROBIAL COMPOUNDS

Following this brief survey of the discovery of the present wide range of antimicrobial compounds, we may now turn to the main theme of the book. We shall be concerned with the biochemical mechanisms that underlie the action of compounds used in the battle against micro-organisms. This topic has a twofold interest. In the long run a detailed understanding of drug action at the molecular level may generate ideas for the design of entirely novel antimicrobial agents. In spite of the great power and success of the drugs currently available, there are still areas where improvements can undoubtedly be made. The other interest in antimicrobial agents is the light that their activity can throw on the subject of biochemistry itself. Antibacterial agents, particularly the antibiotics, often have a highly selective action on biochemical processes. They may block a single reaction within a complex sequence of events. The use of such agents has often revealed details of biochemical processes that would otherwise have been difficult to disentangle. Our present knowledge owes much to experiments with such compounds as puromycin, penicillin and oligomycin. Understanding of the biochemistry of antimicrobial action has been built up slowly and painfully with many false starts and setbacks. Since about 1960, however, a much greater insight into the action of antibacterial compounds has been achieved, and for most of the compounds commonly used in medical practice at least an outline can be given of the biochemical effects underlying their action on bacteria. In contrast, knowledge of the mechanism of action of antiprotozoal drugs, which were discovered long before the antibacterial drugs, has lagged well behind. This is due mainly to the much greater difficulty in isolating and working with protozoa outside the animal body, but interest has also been concentrated on bacteria because of their special importance in infectious disease and because of their widespread use in biochemical and genetic research. Recently, the actions of antifungal and antiviral drugs have attracted increasing attention.

DEVELOPMENT OF BIOCHEMICAL KNOWLEDGE OF ANTIMICROBIAL ACTION

Understanding of how antimicrobial compounds bring about their effects has accumulated gradually. Several levels of progress can be distinguished and will be separately discussed.

Pharmacological biochemistry

When an antimicrobial agent is used systemically its effectiveness is determined by various factors that govern its behaviour in the body. Nowadays the absorption, distribution and excretion of drugs are essential subjects for investigation. Activity requires the maintenance of a sufficient concentration of the drug at the site of action; this concentration must continue long enough to allow the body's defences time to defeat the infection. The concentration attained depends on relative rates of absorption and excretion. It is also affected by metabolic changes brought about by the body's enzymes acting on the drug. In most cases metabolism in the body results in the inactivation of antimicrobial agents, but several examples are known where metabolism is essential for converting an inactive administered compound into an active drug circulating in the body (Chapters 4 and 6). The degree of binding of the drug to various tissues may also be important. Some drugs are firmly bound to plasma proteins. This increases their persistence in the body but may also lower their effectiveness if activity depends on the concentration of free (unbound) compound present in the blood. There is always an equilibrium between free and bound drug but for strongly bound compounds the amount of drug in free solution may be very small. The methods for studying such factors are now well established; it is often convenient to use drugs which carry a radioactive atom as a 'label' at some point in their molecule. Information of this sort is an important preliminary to understanding the therapeutic action of drugs. It can help to explain differences of effects between different species and for a new compound it gives a firm base for recommendations on the size and frequency of doses used in treating human patients.

Nature of the biochemical systems affected

As long as antimicrobial compounds have been known, workers have attempted to explain their action in biochemical terms. Ehrlich made a tentative beginning in this direction when he suggested that the arsenicals might act by combining with thiol groups on the protozoal cells. He was, however, severely limited by the elementary state of biochemistry at that period. By the time the sulphonamides were discovered the biochemistry of small molecules was much more advanced and reasonable biochemical explanations of sulphonamide action were soon available. However, many of the antibiotics which followed presented very different problems. Attempts to apply biochemical methods to the study of their action led to highly conflicting answers. At one stage a count showed that 14 different biochemical systems had been suggested as the site of action of streptomycin against bacteria. Much of this confusion arose from a failure to distinguish between primary and secondary effects. The biochemical processes of bacterial cells are closely interlinked. Hence disturbance of any one important system is likely to have effects on many of the others. Methods had to be developed that would distinguish between the primary biochemical effect of an antimicrobial agent and other changes in metabolism that followed as a consequence. Once these were established more accurate assessments could be made of the real site of action of various antimicrobial compounds. The limiting factor then became the extent of biochemical information about the nature of the target site. From about 1955 onwards there has been a dramatic increase in the understanding of the structure, function and synthesis of macromolecules. Most of the important antibiotics were found to act by interfering with the biosynthesis or function of macromolecules and the development of new techniques has provided the means of locating their site of action with some confidence.

Methods used for the study of the mode of action of antimicrobial compounds

Experience gained over the last twenty five years has helped to evolve reasonably systematic procedures for working out the primary sites of action of many antimicrobial compounds. Once the primary site of action is established the overall effect of a drug on the metabolism of microbial cells can often be explained. Many of the techniques used in elucidating the mode of action of antimicrobial agents are discussed in later chapters but it may be helpful to set them out in a logical sequence.

(1) Where the chemical structure is known, it is studied carefully to determine whether a structural analogy exists with part or whole of a biologically important molecule, i.e. a metabolic intermediate or essential

cofactor, nutrient, etc. An analogy may be immediately obvious, but sometimes it becomes apparent only through imaginative model building or by hindsight when the target site of the compound has been revealed by other means. Analogies of structure can sometimes be misleading and should only be used as a preliminary indication.

(2) The next step is to examine the effects of the compound on the growth kinetics and morphology of suitable target cells. A cytocidal effect shown by reduction in viable count probably indicates damage to the cell membrane. This can be confirmed by observation of leakage of potassium ions, nucleotides or amino acids from the cells. Severe damage leads to cell lysis. Examination of cells by electron microscopy may show morphological changes which indicate interference with the synthesis of one of the components of the cell wall. Many antibiotics have only a cytostatic action and do not cause any detectable morphological changes.

(3) If the antimicrobial agent does not act by damaging the membrane, attempts are made to reverse its action by addition to the medium of various supplements. Nutrients, including oxidizable carbon sources, fatty and amino acids, intermediary metabolites such as purines and pyrimidines, vitamins and bacterial growth are tested in turn. If reversal is achieved, this may point directly to the reaction or reaction sequence which is blocked by the inhibitor. Valuable confirmatory evidence can sometimes be obtained by the use of auxotrophic organisms requiring a compound known to be the next intermediate in a biosynthetic sequence beyond the reaction blocked by the antimicrobial agent. An auxotroph of this type should be resistant to the action of the inhibitor. Inhibition in a biosynthetic sequence may also be revealed by accumulation of the metabolite immediately *before* the blocked reaction. Unfortunately the actions of many antimicrobial agents are not reversed by exogenous compounds. This especially applies to compounds which interfere with the polymerization stages in nucleic acid and protein biosynthesis where reversal is impossible.

(4) The ability of an inhibitor to interfere with the supply and consumption of ATP is usually examined since any disturbance of energy metabolism has profound effects on the biological activity of the cell. The inhibitor is tested against the respiratory and glycolytic activities of the micro-organism, and the ATP content of the cells is measured.

(5) Useful information can often be gained by observing the effect of an antimicrobial agent on the kinetics of uptake of a radiolabelled nutrient, such as glucose, acetate, a fatty acid, an amino acid, phosphate, etc. It is preferable to choose conditions where a steady uptake of nutrient is established. Changes in rate of incorporation which follow the addition of the drug are measured and compared with effect on growth. A prompt interference with incorporation of a particular nutrient may provide a good clue to the primary site of action.

(6) An antimicrobial compound that inhibits protein or nucleic acid synthesis in cells without interfering with (i) membrane function, (ii) the biosynthesis of the immediate precursors of proteins and nucleic acids or (iii) the generation and utilization of ATP, very probably inhibits macromolecular synthesis at the polymerization level. Because of the close interrelationship between protein and nucleic acid synthesis indirect effects of the inhibition of one process on the other are carefully distinguished. For example, a compound that interferes directly with the biosynthesis of RNA indirectly inhibits protein biosynthesis as the supply of messenger RNA is exhausted. Again, inhibitors of protein synthesis eventually arrest DNA synthesis because of the requirement for continued protein biosynthesis for the initiation of new cycles of DNA replication. A study of the kinetics of the inhibition of each macromolecular biosynthesis in intact cells is valuable since indirect inhibitions appear later than direct effects.

(7) When the inhibited biochemical system has been identified in intact cells, more detailed information can then be obtained with cell-free preparations. This may involve the isolation of suspected target enzymes, or cell organelles such as membranes, ribosomes or purified highly polymerized nucleic acids. The antimicrobial compound is tested for inhibitory activity against the suspected target reaction *in vitro*. In this way more insight may be gained about the molecular nature of the interaction between the inhibitor and the target site. There is always a danger, however, of non-specific effects *in vitro* especially at high concentrations of drug. Failure to inhibit the suspected target reaction *in vitro* on the other hand even with very high concentrations of drug cannot rule out inhibition of the same reaction in intact cells for a number of reasons. (i) The drug may be metabolized either by the host or by the micro-organism to an inhibitory derivative. (ii) Extensive purification of an enzyme may desensitize it to the inhibitor by altering an allosteric target. (iii) The site of inhibition in the intact cell may be part of a highly integrated structural assembly which is disrupted during the preparation of a cell-free system again causing a loss of sensitivity to the inhibitor. The use of cell-free preparations from drug-resistant mutants is sometimes useful in the localization of the region of attack. This approach has been ingeniously exploited in identifying the target site of streptomycin in bacterial ribosomes (Chapter 4).

Selectivity of action of antimicrobial agents

In the study of any useful antimicrobial agent it is not sufficient just to explain its action on the metabolism of the micro-organism. The compound must by definition be selective in its action, and a reason must be sought for this selectivity. Thus it is usually necessary to study the actions of these compounds on the biochemistry of the host cell as well as that of the parasite.

The basis of selectivity varies from one drug to another. The process inhibited may occur only in the microbial cell, so there is no reason for the host cells to be affected. Other agents act on biochemical mechanisms found in both microbial and animals cells, but for some reason affect only the former. In such cases the reason for this differential action is still obscure and evidently requires further study. Yet another type of selectivity depends on the concentrating of the antimicrobial agent within the microbial cell, but not in the host cell. Thus even though at the ultimate site of action the drug is equally inhibitory to both, its greater concentration in the micro-organism ensures the necessary selectivity. Here the question shifts to the reason for the selective concentration, and our understanding of these mechanisms at the molecular level is still limited.

The molecular interaction between the antimicrobial agent and the target site

Early mode of action studies usually concentrated on revealing the biochemical basis for the antimicrobial effect. Increasingly scientists are no longer satisfied with this level of explanation alone. There is a wish to understand drug action in molecular terms; ideally the biochemical effects should be interpreted in terms of specific interactions between the drug and its target or receptor molecule. In order to achieve this level of understanding of drug action specialized techniques must be used to reveal the detailed three-dimensional structure of the drug and its receptor. Whilst the former objective is usually readily attainable the structural elucidation of macro-molecular entities and supramolecular organized structures like membranes and ribosomes is a formidable undertaking. Nevertheless the combination of advanced physical techniques and recombinant DNA methodologies provides analytical tools of unprecedented power to probe drug action at the molecular level. The future of chemotherapeutic research can be expected to show a strong bias in this direction.

Meanwhile a start may be made by studying the exact structural requirements for antimicrobial activity. With the complex antibiotics the structures of active molecules are highly specific; minor chemical changes can result in complete inactivation. Only with more simple compounds can the significance of chemical structure be satisfactorily explained in terms of biological activity. A successful antimicrobial suitable for systemic use must have a combination of properties: suitable absorption and distribution in the animal body, ability to penetrate or concentrate in the micro-organism, and selective action on a target site within the cell. Each of these may require some different molecular feature. For optimum activity all these features must be present together in the same molecule and must not interfere with one another. The success of the best antibiotics may be due to contributions by different parts of the molecules to the overall pattern of action. Biochemical research on isolated systems may eventually distinguish the contributions of these different components.

The biochemistry of microbial resistance

The therapeutic value of an antimicrobial agent often declines after prolonged use through the emergence of organisms which are no longer sensitive to the compound. This problem of expanding practical importance has been studied by microbiological, biochemical and molecular genetic methods. Such studies will usually show the means by which resistance has been acquired, either by selection or by genetic transfer through a plasmid or by phage infection. They will also define the biochemical changes that render an organism resistant. The results of such studies are discussed in Chapter 8. The mechanism of some forms of resistance still needs further study and there is obviously great practical interest in methods of preventing the acquisition of resistance or of combating resistance once it has arisen.

The practical value of resistance studies has already been proved by the development of successful antibacterials in which the original molecules have been chemically modified to render them insusceptible to bacterial enzymes responsible for important types of resistance.

SCOPE AND LAYOUT OF THE PRESENT BOOK

In this book we have tried to select well-established evidence for the biochemical action of many of the best-known agents used in medicine. Those compounds where the mode of action is in considerable doubt receive only passing mention. Some antimicrobial compounds of particular biochemical significance are also included even though they have no therapeutic value. In one chapter we have brought together information on the action of antifungal agents including some which are used against plant pathogens. In the same chapter several compounds active against protozoa are discussed. Compounds showing activity against human virus infections including compounds active against the AIDS virus are also considered.

We have classified drugs mostly according to their types of biochemical action rather than on grounds of chemical structure. The last two chapters consider the mechanisms by which antimicrobial compounds enter the target cells, and the biochemistry of the development of resistance to antibacterial agents.

FURTHER READING

Cohen, S. S. (1979) Comparative biochemistry and drug design for infectious disease. *Science*, **205**, 964.

Gale, E. F., Cundliffe, E., Reynolds, P. E., Richmond, M. H. and Waring, M. J. (1981) *The Molecular Basis of Antibiotic Action*, 2nd edn, John Wiley & Sons.

Garrod, L. P., Lambert, H. P. and O'Grady, L. (1973) *Antibiotic and Chemotherapy*, 4th edn., Churchill Livingstone, Edinburgh.

Glasby, J. S. (1979) *Encyclopaedia of Antibiotics*, 2nd edn., Wiley.

Hahn, F. E. (ed.) (1979) *Antibiotics*, Vol. 5, Part 1, *Mechanism of Action of Antibacterial Agents*; Part 2, *Mechanism of Action of Antieukaryotic and Antiviral Compounds*, Springer Verlag.

Hugo, W. B. and Russell, A. D. (eds) (1987) *Pharmaceutical Microbiology*, 4th edn., Blackwell, Oxford.

Korzybski, T., Kowszyk-Gindifer, Z. and Kurylowicz, W. (1979) *Antibiotics, Origin, Nature and Properties* (3 vols.), English edn., American Society for Microbiology.

Pratt, W. B. (1973) *Fundamentals of Chemotherapy*, Oxford University Press, Oxford.

Sneader, W. (1985) *Drug Discovery: The Evolution of Modern Medicines*, John Wiley & Sons.

Umezawa, H. (ed.), *Index of Antibiotics from Actinomycetes*, Vol. 1, 1967, Vol. 2, 1978, Japanese Scientific Societies Press.

2 The bacterial cell wall – a vulnerable shield

FUNCTIONS OF THE WALL

In the search for differences between bacteria and animal cells that could provide the basis for a selective antibacterial attack, one evident distinction lies in their general structure. The animal cell is relatively large and has a complex organization; its biochemical processes are compartmentalized and different functions are served by the nucleus with its surrounding membrane, by the mitochondria and by various other organelles. The outer membrane is thin and lacks rigidity. The cell exists in an environment controlled in temperature and in osmolarity and it is constantly supplied with nutrients from the extracellular fluid bathing it. The smaller bacterial cell lives in a variable and often unfavourable environment. In particular it must be able to withstand considerable changes in osmolarity. Some bacteria have relatively high concentrations of solutes of low molecular weight in their cytoplasm. Such a cell suspended in water or in dilute solutions will develop a high internal osmotic presure. This would inevitably disrupt its membrane unless it were provided with a tough, rigid outer coat. This coat is the cell wall, a characteristic bacterial structure entirely lacking in animal cells. It has a protective function for the bacterium, but at the same time it is vulnerable to attack, and a number of effective antibacterials owe their action to their ability to disturb the processes by which the wall is synthesized. Since there is no parallel biosynthetic mechanism in the animal cell, substances affecting this process may have quite a low toxicity.

The term 'wall' will be used to describe all the cell covering which lies outside the cytoplasmic membrane. 'Envelope' comprises both the wall and the cytoplasmic membrane.

WALL STRUCTURE

Bacterial wall structure varies considerably from one species to another, but follows general patterns which are related to the broad morphological classification of bacteria. Classically this has been based on behaviour towards the Gram stain, but the well-tried division into Gram-positive and Gram-negative types has a significance far beyond that of an empirical staining reaction. The most evident differences are worth recalling.

Many Gram-negative bacteria are highly adaptable organisms which will use inorganic nitrogen compounds, mineral salts and a simple carbon source for synthesis of their whole structure. Their cytoplasm has a relatively low osmolarity. The typical Gram-positive coccus or bacillus tends to be more exacting in its nutritional needs. Such organisms lack some synthetic abilities and require various amino acids, vitamins and accessory factors for growth. They are usually cultivated on rich undefined broths or on fairly elaborate synthetic media. They concentrate amino acids, nucleotides and other metabolites of low molecular weight in their cytoplasm and consequently have a high internal osmolarity. Bacteria do not all fit this neat division. The Gram-negative cocci, the rickettsias, the chlamydias and the spirochaetes, for example, are all Gram-negative bacteria with exacting growth requirements. The mycoplasmas lack a rigid wall structure and, although technically Gram-negative, are best treated as a separate group lying outside the usual Gram stain classification.

Most of the work on wall structure has been done with Gram-positive cocci and bacilli and with enteric bacteria and other Gram-negative rods. The extent to which the structural generalizations apply to groups outside these classes is uncertain.

The Gram-positive wall

Many Gram-positive bacteria have relatively simple walls. The wall, which lies above the cell membrane, is usually beween 15 and 50 nm thick. For biochemical studies, bacteria are broken by shaking a suspension with small glass beads; the walls are separated from cytoplasmic material by washing and differential centrifugation. In electron micrographs these wall preparations resemble empty envelopes, torn in places where the cytoplasmic contents were released. With Gram-positive bacteria the wall is obtained relatively free from other bacterial components, but Gram-negative wall preparations (Fig. 2.1) usually contain some material from the cytoplasmic membrane.

The major part of the Gram-positive wall is a large polymer comprising two covalently linked components. One of these components, forming at least 50% of the wall mass, is peptidoglycan (sometimes referred to as murein or mucopeptide). Its cross-linked structure provides a tough fibrous fabric giving strength and shape to the cell and enabling it to withstand a high internal osmotic pressure. The amount of peptidoglycan in the wall shows that it covers the cell in a multilayered structure, with cross-linking both within and between the layers. Attached to the peptidoglycan is an acidic

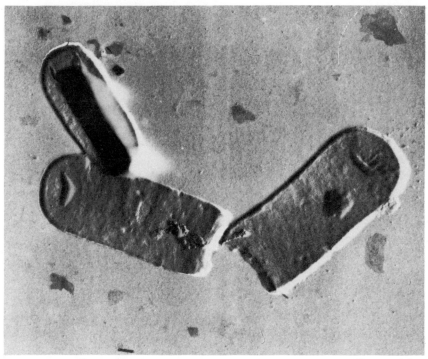

Figure 2.1 Electron micrograph of isolated cell envelopes from *Pseudomonas aeruginosa* shadowed with metal. We thank Mr A. Davies and Mrs M. Bentley for this picture.

polymer, accounting for 30–40% of the wall mass, which differs from species to species. Often this is a teichoic acid – a substituted poly(D-ribitol 5-phosphate) (see Fig. 2.8) – or a substituted glycerol 3-phosphate. In some bacteria teichoic acid is replaced by poly(*N*-acetylglucosamine l-phosphate) or teichuronic acid (a polymer containing uronic acid and *N*-acetylhexosamine units). Bacteria that normally incorporate teichoic acid in their walls can switch to teichuronic acid under conditions of phosphate limitation. The acidic character of the polymer attached to the peptidoglycan ensures that the cell surface is strongly polar and carries a negative charge. This may influence the passage of ions, particularly Mg^{2+}, into the cell.

The teichoic acid or other acidic polymer is readily solubilized and released from the insoluble peptidoglycan by hydrolysis in cold acid or alkali. The nature of the linkage is described later.

Other components of the Gram-positive wall vary widely from species to species. Protein is often present to the extent of 5–10%, and protein A of *Staphylococcus aureus* is apparently linked covalently to peptidoglycan. Proteins and polysaccharides frequently occur in the outermost layers and provide the main source of the antigenic properties of these bacteria. Mycobacteria and a few related genera differ from other Gram-positive bacteria in having large amounts of complex lipids in their wall structure.

The Gram-negative wall

The Gram-negative wall is far more complex. Wide-ranging studies of its structure have been concentrated on the Enterobacteriaceae and on *Escherichia coli* in particular.

When cells of *Escherichia coli* are fixed, stained with suitable metal salts, sectioned and examined by electron microscopy, the cytoplasmic membrane can be readily identified by its usual 'sandwich' appearance of two electron-dense layers separated by a lighter space. Above this, the cell wall appears as a structure containing three electron-dense layers separated by clear layers (Fig. 2.2). The clear layer immediately outside the cytoplasmic membrane is described as the periplasmic space. Here are found soluble enzymes and other components which can be released by submitting the cell to transient non-lethal osmotic shock. The electron-dense layer, about 2 nm thick, immediately outside the periplasmic space represents the peptidoglycan component of the wall. It is much thinner than in Gram-positive bacteria and may constitute only 5% or less of the wall mass. Even so it contributes substantially to wall strength. Cells rapidly lyse when treated with lysozyme, an enzyme which specifically degrades peptidoglycan. In *Escherichia coli* the peptidoglycan is covalently linked to a lipoprotein which probably projects into the outer regions of the wall. The outer regions of the Gram-negative cell wall have been the most difficult to characterize. The various components together form a structure 6–10 nm thick, called the outer membrane. Like the cytoplasmic membrane it is basically a lipid bilayer, hydrophobic in the interior with hydrophilic groups at the outer surfaces. It also has protein components which penetrate the layer partly or completely and form the membrane 'mosaic'. Despite these broad structural similarities, the outer membrane differs widely in composition and function from the cytoplasmic membrane. Its main constituents are lipopolysaccharide, phospholipids, fatty acids and proteins. The phospholipids, mainly phosphatidylethanolamine and phosphatidylglycerol, resemble those in the cytoplasmic membrane. The structure of the lipopolysaccharide is complex and varies considerably from one bacterial strain to another. The molecule has three parts (see Fig. 2.3). The core is built from 3-deoxy-D-*manno*-octulosonic acid (KDO), hexoses, heptoses, ethanolamine and phosphoric acid as structural components. The three KDO residues contribute a structural unit which strongly binds the divalent ions of magnesium and calcium, an important feature stabilizing the membrane. Removal of these ions by chelating agents

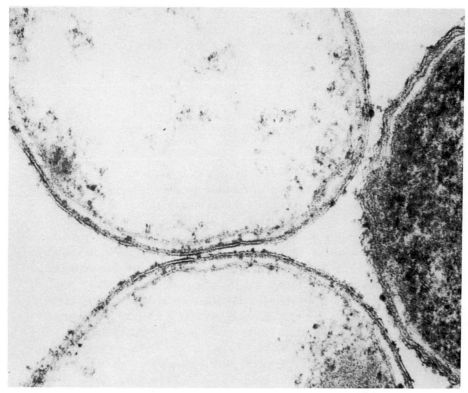

Figure 2.2 Electron micrograph of cross-section of parts of *Escherichia coli* cells showing the multiple structure of cell wall and cell membrane. We thank Mr A. Davies and Mrs M. Bentley for this picture.

leads to release of some of the lipopolysaccharide into the medium; at the same time the membrane becomes permeable to compounds that would otherwise be excluded. The core polysaccharide is linked to the antigenic side chain, a polysaccharide which can vary greatly from one strain to another even within the same bacterial species. Usually it comprises about 30 sugar units, although these can vary both in number and in structure. It forms the outermost layer of the cell and is the main source of its antigenic characteristics. At the opposite end, the core of the lipopolysaccharide is attached to a moiety known as lipid A which can be hydrolysed to glucosamine, long-chain fatty acids, phosphate and ethanolamine. The fatty acid chains of lipid A, along with those of the phospholipids, align themselves to form the hydrophobic interior of the membrane. The outer membrane is asymmetric with lipopolysaccharide exclusively on the outer surface and phospholipid mainly on the inner surface.

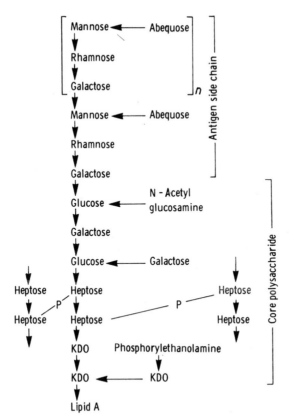

Figure 2.3 Tentative structure of the lipopolysaccharide of the cell envelope of *Salmonella typhimurium*. The diagram has been simplified by omitting the configuration of the glycosidic linkages and omitting the *O*-acetyl groups from the abequose units. KDO, 3-deoxy-D-*manno*-octulosonic acid. Lipid A consists of a β-1,6-linked diglucosamine residue to which lauric, myristic, palmitic and 3-D(−)-hydroxymyristic acids are bound. The heptose residues of three lipopolysaccharide polymers are shown linked by phosphate diester bridges although an alternative view is that the polymers may be interlinked by pyrophosphate bridges between the diglucosamine residues of lipid A. Although there are considerable structural variations in the antigen side chain among *Salmonella* species, the core polysaccharide and lipid A are probably common to all wild-type *Salmonella*. The core structure in *Escherichia coli* is more variable.

Many proteins have been identified as components of the outer membrane. The most abundant in *Escherichia coli* are the matrix protein and lipoprotein. Matrix protein, also known as protein 1, has been fully sequenced. Its molecular weight is 37 205. Together with other rather similar proteins it forms a hexagonal lattice in the outer membrane with a 7.7 nm repeating unit. This pattern is best seen by electron microscopy of spheroplasts lacking peptidoglycan. Such examination also reveals triplets of indentations in the membrane surface each 2 nm in diameter, and 3 nm apart, through which the stain used in the preparation readily penetrates. This is interpreted as showing that the matrix protein molecules stretch across the membrane in groups of three, enclosing pores through which water and small molecules can pass. Such proteins have therefore been called porins. The size of the pores explains the selective permeability of the bacterial wall, freely allowing the entry of hydrophilic molecules up to a maximum molecular weight of 600–700. Larger flexible molecules may also diffuse through the pores although with more difficulty. Artificial vesicles can be made with outer membrane lipids. Without protein these are impermeable to solutes, but when porins are incorporated they show permeability characteristics similar to those of the outer membrane itself.

Lipoprotein is another major component of the outer membrane proteins. About one-third is linked to peptidoglycan and the remaining two-thirds are unattached, but form part of the membrane. The nature of the attachment of lipoprotein to the side chains of peptidoglycan is discussed later. About one in 12 of the peptide side chains is thus substituted. The arrangement provides a structural anchorage of the outer membrane to the peptidoglycan layer. The fatty acid chains of the lipoprotein presumably align themselves in the hydrophobic inner layer of the outer membrane and the protein moiety may possibly associate with matrix protein, reinforcing the pore structure.

Many other proteins with specialized functions have been identified in the outer membrane. Some of these are transport proteins, allowing access to molecules such as vitamin B_{12} or nucleosides which are too large to penetrate the pores of the membrane. These proteins also possess receptor sites which are recognized by specific phages which attack the bacteria.

STRUCTURE AND BIOSYNTHESIS OF PEPTIDOGLYCAN

The structure and biosynthesis of peptidoglycan have special significance relative to the action of a number of important antibacterials and have been studied extensively.

The biosynthesis of peptidoglycan was first worked out with *Staphylococcus aureus*. Although bacteria show many variations in peptidoglycan structure, the biosynthetic sequence in *Staphylococcus aureus* will serve to illustrate the general features of the process.

Cell walls from *Staphylococcus aureus* were gently hydrolysed with trichloro-acetic acid to remove the teichoic acid. The insoluble peptidoglycan was then hydrolysed with hot mineral acid. The hydrolysate contained amino sugars and the four amino acids, glycine, alanine, glutamic acid and lysine in the ratio 5:2:1:1. A clue to the structure and to the early stages of biosynthesis was an observation by Park that subinhibitory concentrations of penicil-

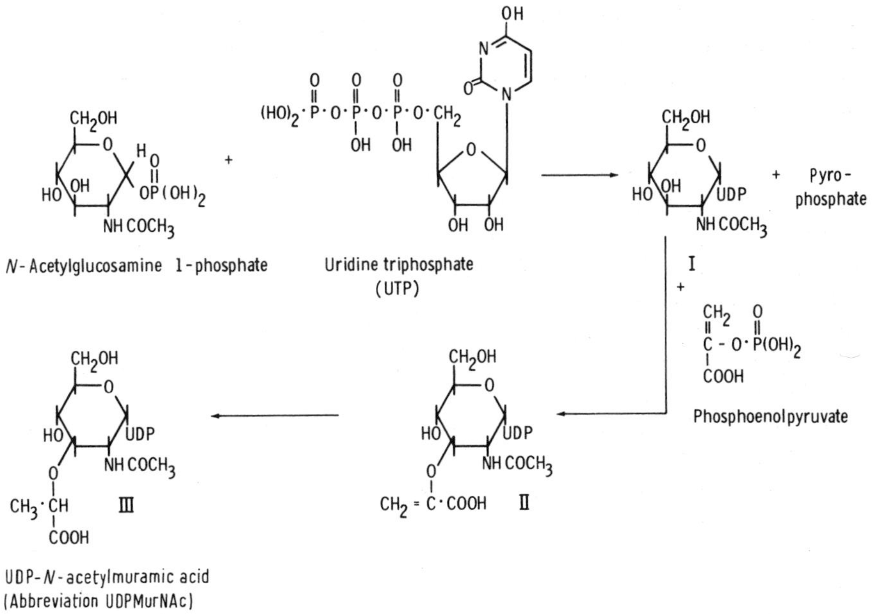

Figure 2.4 Peptidoglycan synthesis in *Staphylococcus aureus*. Stage 1: formation of UDP-N-acetylmuramic acid.

lin caused the accumulation of three nucleotides. He determined their main features, and it is now known that these compounds were III (Fig. 2.4), IV (Fig. 2.5) and the intermediate between these two in which only L-alanine had been attached. The significance of these compounds was not recognized at the time, but later, when it became apparent that penicillin probably inhibits cell wall biosynthesis, the similarity between the hydrolysis products of peptidoglycan and of the Park nucleotides was evident. From this and other experiments the peptidoglycan structure was gradually pieced together and the details of its biosynthesis were worked out, mainly by Strominger and Tipper. The biosynthesis may be conveniently divided into four stages.

Stage 1.   Synthesis of UDP-N-acetylmuramic acid

The biosynthesis starts with two products from the normal metabolic pool, N-acetylglucosamine 1-phosphate and UTP (Fig. 2.4). UDP-N-acetyl-glucosamine (I) is formed by the usual elimination of pyrophosphate. This nucleotide reacts with phosphoenolpyruvate by means of a specific trans-ferase to give the corresponding 3-enolpyruvyl ether (II). The pyruvyl group is then converted to lactyl by a reductase requiring NADPH, the

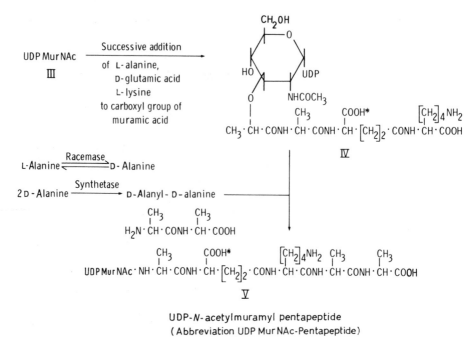

UDP-N-acetylmuramyl pentapeptide
(Abbreviation UDP Mur NAc-Pentapeptide)

Figure 2.5 Peptidoglycan synthesis. Stage 2: formation of UDP-N-acetylmuramyl pentapeptide. Addition of each amino acid and the final dipeptide requires ATP and a specific enzyme. L-Lysine is added to the γ-carboxyl group of D-glutamic acid; the α-carboxyl group (marked *) is amidated at a later stage in the biosynthesis.

product being UDP-N-acetylmuramic acid (III, UDPMurNAc). Muramic acid (3-O-D-lactyl-D-glucosamine) is a distinctive amino sugar derivative found only in the peptidoglycan of cell walls.

Stage 2. Building the pentapeptide side chain

Five amino acid residues are next added to the carboxyl group of the muramic acid nucleotide (Fig. 2.5). Each step requires ATP and a specific enzyme. L-Alanine is added first. The next two residues are D-glutamic acid and then L-lysine. The lysine however is attached through its α-amino group to the γ-carboxyl group of the glutamic acid. The α-carboxyl group of the glutamic acid is amidated at a later stage in the biosynthesis, so the second amino acid residue is sometimes referred to as D-*iso*glutamine. The biosynthesis of the pentapeptide is completed by addition not of an amino acid but of a dipeptide, D-alanyl-D-alanine, which is separately synthesized. A racemase acting on L-alanine gives D-alanine, and a synthetase then joins two molecules giving the dipeptide. The completed UDP-N-acetylmuramyl intermediate (V) with its pendant peptide group will be referred to as the 'nucleotide pentapeptide'.

Stage 3. Membrane-bound reactions leading to a linear peptidoglycan polymer

The biosynthesis up to this point involves relatively small molecules, is cytoplasmic and the reactions are well characterized. The succeeding steps occur on membrane structures and are therefore less certainly defined. Although the course of each individual reaction is well established, the reaction sequence and the mode of attachment of newly synthesized units to existing wall material is still uncertain.

The first step involves the formation of a pyrophosphate link between the nucleotide pentapeptide and undecaprenyl phosphate (the phosphate ester of a C_{55} isoprenoid alcohol) which is a component of the cytoplasmic membrane (Fig. 2.6). In this reaction UMP is released and becomes available for reconversion to UTP which is needed in the first step of peptidoglycan biosynthesis (Fig. 2.4). All subsequent reactions occurring while the intermediates are linked to undecaprenyl phosphate take place without release from the membrane. An essential step in this membrane-bound reaction sequence is the addition of a second hexosamine residue through a typical glycosidation by UDP-N-acetylglucosamine (Fig. 2.6). The disaccharide (VII) is formed by a 1-4β linkage with liberation of UDP. The involvement of undecaprenyl phosphate is not unique to peptidoglycan biosynthesis. It is also concerned in the biosynthesis of the polysaccharide chain in the O-antigen produced by *Salmonella typhimurium* and in the formation of the polysaccharide elements of the lipopolysaccharides of Gram-negative bacteria; in Gram-positive bacteria it fulfils a similar role in the biosynthesis of teichoic acid or polysaccharides of the wall.

At about this point in the biosynthesis of *Staphylococcus aureus* peptidoglycan an extending group is added to the ε-amino group of the lysine unit in the nucleotide pentapeptide. Glycine and a glycine-specific transfer RNA are involved in this process during which a pentaglycine group is added. The reaction, unlike the tRNA reactions in protein biosynthesis, does not require the presence of ribosomes; the five glycine units are added successively to the lysine from the nitrogen end (the reverse direction to protein biosynthesis). Since the resultant product (VIII) has 10 amino acid units it will be referred to as the disaccharide decapeptide. This still has a free terminal amino group. In the biosynthesis of peptidoglycans from certain other bacterial species no extending group is added. Later reactions then involve the ε-amino group of lysine (or equivalent diamino acid) instead of the terminal amino group of glycine. Also during the membrane-bound stage in the biosynthesis of *Staphylococcus aureus* peptidoglycan the carboxyl group of

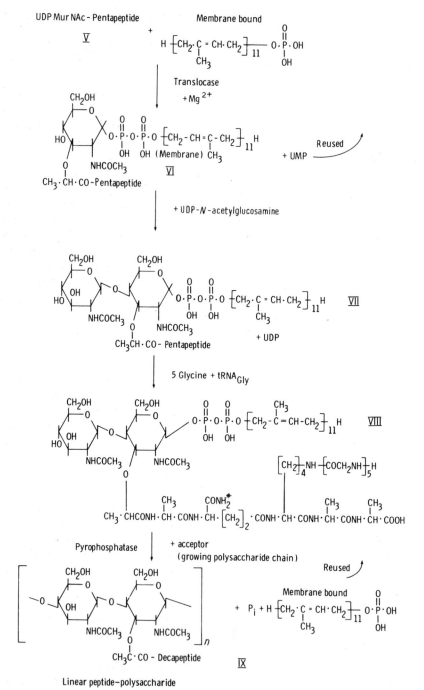

Figure 2.6 Peptidoglycan synthesis. Stage 3: formation of the linear peptidoglycan. The structure of the decapeptide side chain is shown in VIII; the point at which the amide group* is added has not yet been established.

D-glutamic acid is amidated by a reaction with ammonia and ATP. The exact sequence of these three reactions, addition of *N*-acetylglucosamine, chain extension and amidation is uncertain.

The final reaction in this stage is the attachment of the disaccharide decapeptide (VIII) to an undefined 'acceptor' present in the particulate preparations that have been used for investigating these reactions. This acceptor is usually regarded as the growing linear polymer chain. In this reaction the disaccharide with its decapeptide side chain forms a β-linkage from the 1 position of the *N*-acetylmuramic acid residue to the 4-hydroxyl group of the terminal *N*-acetylglucosamine residue in the growing polysaccharide chain. The membrane lipid is left as undecaprenyl pyrophosphate and is reconverted by a specific pyrophosphatase to the corresponding phosphate ready for another cycle of the membrane-bound part of the synthesis. The growth of the glycan chains thus occurs by successive addition of disaccharide units. The pyrophosphatase reaction is inhibited by bacitracin, an effective anti-Gram-positive antibiotic. The clinical use of this agent is limited to topical infections because of toxicological side effects when used systemically.

Stage 4. Cross-linking

The linear peptidoglycan (IX) formed in stage 3 contains many polar groups which make it soluble in water. It lacks rigidity and toughness. These attributes are introduced by cross-linking, a process well known in the plastics industry for producing similar results in synthetic linear polymers. The reactions leading to the linear peptidoglycan are intracellular or are intimately associated with the cytoplasmic membrane. The energy linked synthetic steps are brought about by conventional biochemical reactions. However, the final stage of peptidoglycan formation when the rigid cross-linked structure is laid down must take place outside the cell membrane, since the material once formed is insoluble and could not be moved from one site to another. The mechanism involved in cross-linking seems especially adapted to this situation; it is a transpeptidation reaction requiring no external supply of ATP or similar compounds. The transpeptidation occurs between the terminal amino group of the pentaglycine side chain and the peptide amino group of the terminal D-alanine residue of another peptide side chain; D-alanine is eliminated and a peptide bond is formed (Fig. 2.7). The loss of D-alanine in this reaction explains a feature that originally caused difficulty in understanding the biosynthetic process. The ratio of alanine (D and L) to glutamic acid or to lysine in *Staphylococcus aureus* peptidoglycan was well established as about 2:1, whilst in the 'pentapeptide', which was the apparent precursor, the ratio was 3:1. The transpeptidation process explains this apparent anomaly. In *Staphylococcus aureus* peptidoglycan the cross-

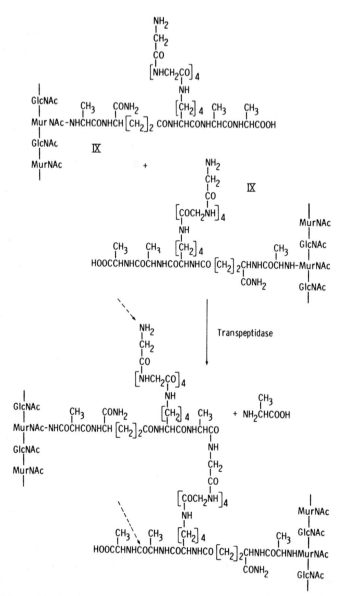

Figure 2.7 Peptidoglycan synthesis. Stage 4: cross-linking of two linear peptidoglycan chains. The linear polymers have the structure IX (Fig. 2.6). GlcNAc, N-acetylglucosaminyl residue. Broken arrows show points at which further cross-links may be formed with other polymer chains.

linking is quite extensive and up to 10 peptide side chains may be bound together by bridging groups. Since the linear polymers themselves are very large it is likely that the whole of the peptidoglycan in a Gram-positive bacterium is made up of units covalently bound together. This gigantic bag-shaped molecule has been called a 'sacculus'. However, even if this extensive covalent bonding really occurs there must be a mechanism for constantly breaking it down and re-forming it to allow for cell growth and division. Peptidoglycan hydrolases which hydrolyse the polysaccharide chains of peptidoglycan and others attacking the peptide cross-links have been isolated from bacteria and probably exert this essential catabolic function during cell growth.

The cross-linking step has been difficult to study because the transpeptidases are found in particulate fractions from cell envelopes and methods have only recently been found for releasing them without inactivation. Progress made with soluble extracellular enzymes and with low-molecular-weight substrates will be discussed later.

Variations in peptidoglycan structure

Many variations are found in peptidoglycan structure between one species of bacteria and another, or even between strains of the same species. Only a general account is possible here. Fuller details will be found in research papers listed at the end of the chapter. All peptidoglycans have the same glycan chain as in *Staphylococcus aureus* except that the glucosamine residues are sometimes N-acylated with a group other than acetyl. The peptide side chains always have four amino acid units alternating L-, D-, L-, D- in configuration. The second residue is always D-glutamic acid, linked through its γ-carboxyl group, and the fourth always D-alanine. The peptidoglycan from *Staphylococcus aureus*, classified by Strominger as Type A2, is characteristic of many cocci. Peptidoglycans of this group, and the related Types A3 and A4, have similar tetrapeptide side chains but vary in their bridging groups. The amino acids in the bridge are usually glycine, alanine, serine or threonine and the number of residues can vary from one to five. In Type A1 peptidoglycans the L-lysine of the Type II peptide side chain is usually replaced by *meso*-2,6-diaminopimelic acid, and there is no bridging group. Cross-linking occurs between the D-alanine of one side chain and the 6-amino group of the diaminopimelic acid of another. This peptidoglycan type is characteristic of many rod-shaped bacteria, both the large family of Gram-negative rods and the Gram-positive bacilli. In the less common Type B peptidoglycans cross-linkage occurs between the α-carboxyl group of the D-glutamic acid of one peptide side chain and the D-alanine of another, through a bridge containing a basic amino acid.

Cross-linking in Gram-negative bacteria

In contrast to the multiple random cross-linkage of peptidoglycan which is found in the Gram-positive cocci, the peptidoglycan of *Escherichia coli* and similar Gram-negative rods shows on average only a single cross-link between one peptide side chain and another. These bacteria contain, besides the transpeptidases concerned in cross-linkage, other enzymes known as DD-carboxypeptidases which specifically remove D-alanine from a pentapeptide side chain. Carboxypeptidase I is specific for the terminal D-alanine of the pentapeptide side chain, whilst carboxypeptidase II acts on the D-alanine at position 4 after the terminal D-alanine has been removed. DD-Carboxypeptidase I seems in some way to limit the extent of cross-linking.

Because of its relatively rigid structure, discussed above, the peptidoglycan sacculus determines the overall shape of the bacterium. Recent evidence suggests that peptidoglycan is laid down with a definite orientation, in which the polysaccharide chains run perpendicular to the main axis of rod-shaped organisms (e.g. *Escherichia coli*).

Attachments to peptidoglycans

Within the cell wall the polymeric peptidoglycan is usually only part of a larger polymer. In Gram-positive cocci it is linked to an acidic polymer, often a teichoic acid (Fig. 2.8).

The point of attachment is through the 6-hydroxyl group of muramic acid

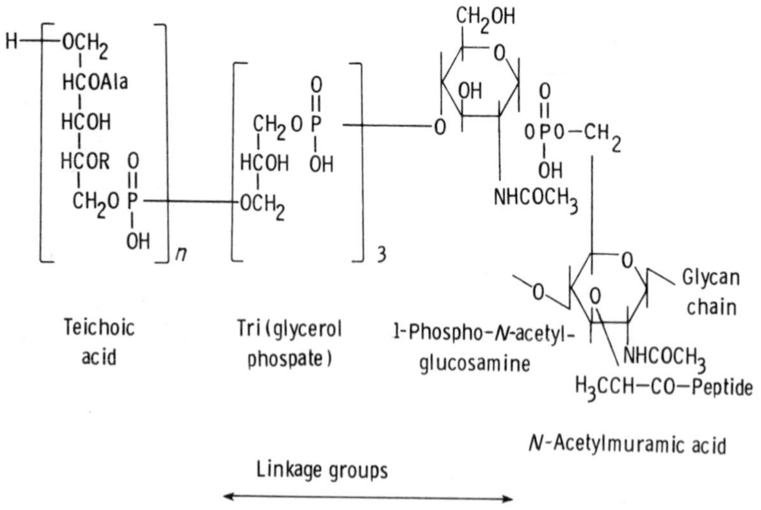

Figure 2.8 Teichoic acid and its linkage to peptidoglycan in the wall of *Staphylococcus aureus*.

in the glycan chain. Only a small fraction of the muramic acid residues are thus substituted. In *Staphylococcus aureus* cell walls teichoic acid is joined to peptidoglycan by a linking unit comprising three glycerol 1-phosphate units attached to the 4-position of *N*-acetylglucosamine which engages through a phosphodiester group at position 1 with the 6-hydroxyl group of muramic acid. This type of linkage seems to occur with polymers other than teichoic acid, e.g. with poly(*N*-acetylglucosamine 1-phosphate) in a *Micrococcus* species. The acid-labile *N*-acetylglucosamine 1-phosphate linkage and the alkali-labile phosphodiester linkage at position 4 explain the ease with which teichoic acid can be split off from peptidoglycan. Within the cell wall the synthesis of teichoic acid is closely associated with that of peptidoglycan.

In the Gram-positive mycobacteria the peptidoglycan carries quite a different polymeric attachment. Arabinogalactan is attached to the 6-position of some of the *N*-glycolylmuramic acid residues of the glycan chain through a phosphate ester group. Mycolic acids (complex, very long-chain fatty acids) are attached by ester links to the C-5 position of arabinose residues of the arabinogalactan. The wall thus has a high lipid content.

In *Escherichia coli* and related bacteria the peptidoglycan carries a lipoprotein as a substituent (Fig. 2.9). The lipoprotein consists of a polypeptide chain of 58 amino acid units of known sequence with lysine at the *C*-terminus and cysteine at the *N*-terminus. This is attached to the 2-carboxyl group of *meso*-2,6-diaminopimelic acid in a peptide side chain of *Escherichia coli* peptidoglycan which has lost both D-alanine groups. Attachment is by an amide link with the ε-amino group in the terminal lysine of the polypeptide. At the opposite end of the polypeptide chain the cysteine amino group carries a long-chain fatty acid joined as an amide, and its sulphur atom forms a thioether link with a long-chain diacylglycerol.

Lipoprotein occurs in enteric bacteria other than *Escherichia coli*, but it may

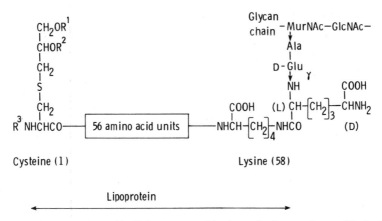

Cysteine (1)   Lysine (58)

Lipoprotein

Figure 2.9 Lipoprotein and its linkage to peptidoglycan in the envelope of *Escherichia coli*.

not be common to all Gram-negative bacteria. Small amounts have been detected with some difficulty in *Proteus mirabilis* and it could not be found in *Pseudomonas fluorescens*.

ANTIBIOTICS THAT INHIBIT PEPTIDOGLYCAN BIOSYNTHESIS

The conclusion that a particular antibiotic owes its antibacterial activity to interference with peptidoglycan biosynthesis rests on a number of lines of evidence.

(a) Bacteria suspended in a medium of high osmotic pressure are protected from concentrations of the antibiotic that would cause lysis and death in a normal medium. Under these conditions the cells lose the shape-determining action of the peptidoglycan and assume a spherical shape; they are then known as spheroplasts. These retain an undamaged cytoplasmic membrane but their wall is deficient or considerably modified. Spheroplasts are in principle viable and if the antibiotic is removed they can divide and produce progeny having normal walls.

(b) A number of species of bacteria lack walls containing peptidoglycan. These include the mycoplasmas, the halophilic bacteria tolerant of high salt concentrations and bacteria in the L-phase, where the normal wall structure is greatly modified. If a compound inhibits the growth of common bacteria but fails to affect bacteria of these special types, it probably owes its activity to interference with peptidoglycan synthesis.

(c) Subinhibitory concentrations of these antibiotics often cause accumulation in the culture medium of uridine nucleotides of *N*-acetylmuramic acid with varying numbers of amino acid residues attached. These compounds represent intermediates in the early stages of peptidoglycan biosynthesis. When an antibiotic causes a block at an early point in the reaction sequence it is not surprising to find accumulation of the intermediate immediately preceding the block. However, quantities of muramic acid nucleotides are also found in bacteria treated with antibiotics that are known to affect later stages in peptidoglycan biosynthesis. It seems that all the biosynthetic steps associated with the membrane are closely interlocked, and inhibition of any one of them leads to accumulation of the last water-soluble precursor, UDP-*N*-acetylmuramyl pentapeptide (V, Fig. 2.5).

Individual antibiotics having this type of action will now be considered.

Bacitracin

Bacitracin, a polypeptide antibiotic (Fig. 2.10), is too toxic for systemic treatment but is sometimes used locally to kill Gram-positive bacteria, e.g. in surgery of the colon. Its effect on peptidoglycan biosynthesis depends upon

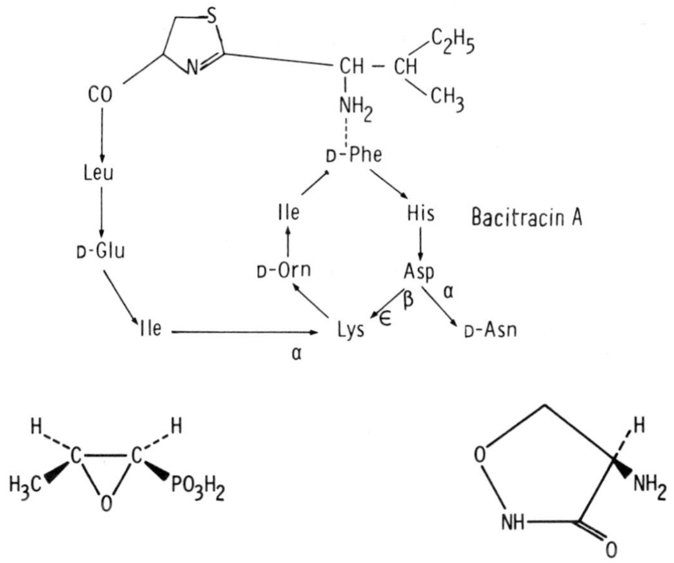

Figure 2.10 Antibiotics affecting intracellular stages in the biosynthesis of peptido-glycan.

its ability to bind specifically to polyprenyl pyrophosphates in the presence of magnesium ions. In the formation of the linear peptidoglycan (IX, Fig. 2.6) the membrane-bound undecaprenyl pyrophosphate is released. Normally this is converted by a pyrophosphatase to the corresponding phosphate which thus becomes available for reaction with another molecule of UDPMurNAc-pentapeptide (V). Complex formation of the pyrophosphate with bacitracin blocks this process and so eventually halts the synthesis of peptidoglycan. Bacitracin also inhibits sterol biosynthesis in animal tissues through complex formation with intermediates such as farnesyl pyrophosphate.

Fosfomycin (phosphonomycin)

This antibiotic has the remarkably simple structure shown in Fig. 2.10. It acts on infections caused by both Gram-positive and Gram-negative bacteria but although its toxicity seems to be low it has achieved only limited use in clinical practice. Its inhibitory action is exerted on the first step of peptido-glycan biosynthesis, namely the condensation of UDP-N-acetylglucosamine (I) with phosphoenolpyruvate by means of a transferase giving the in-termediate (II) that subsequently yields UDP-N-acetylmuramic acid (III) on reduction (Fig. 2.4). This is demonstrated by tests on the isolated enzyme system. Moreover the inhibitory action of fosfomycin towards *Enterobacter cloacae* can be reversed by the UDP-muramyl tripeptide or pentapeptide intermediates characteristic of that organism, indicating that the block must be at an earlier step in the biosynthesis.

Cycloserine

This antibiotic also has a simple structure (Fig. 2.10). It is active against a number of bacterial species but has found little clinical use because of the central nervous system disturbances which are sometimes experienced by patients. Cycloserine shows the usual effects that characterize compounds acting on peptidoglycan biosynthesis, and when cultures of *Staphylococcus aureus* are grown with subinhibitory concentrations of cycloserine considerable amounts of the peptidoglycan precursor (IV) (Fig. 2.5) accumulate in the medium. This suggests a blockage in the biosynthesis immediately beyond this point.

Examination of its effects on enzymes present in cell wall extracts showed cycloserine to inhibit both alanine racemase and D-alanyl-D-alanine synthetase, the two enzymes concerned in making the dipeptide for completion of the pentapeptide side chain. Molecular models reveal that cycloserine is structurally related to one possible conformation of D-alanine, so its inhibitory action on these enzymes appears to be a classical example of isosteric interference.

The observation that the action of cycloserine is specifically antagonized by the addition of D-alanine to the growth medium supported the postulated site of action. Tests on the synthetase enzymes isolated from different bacterial species showed that the affinity of cycloserine for the synthetase was much greater than that of the natural substrate, the ratio of K_m to K_i being about 100.

In a compound acting purely as a competitive enzyme inhibitor, this sort of K_m/K_i ratio is probably essential for the exhibition of useful antibacterial activity. The greater affinity of cycloserine for the enzyme may be connected with its rigid structure. This could permit a particularly accurate fit to the active centre of the enzyme, either in the state existing when the enzyme is uncombined with its substrate or in some modified conformation which is assumed during the normal enzymic reaction. Rigid structures of narrow molecular specificity are common among antimicrobial agents and similar considerations may apply to other types of action; this theme will recur in later sections.

Cycloserine enters the bacterial cell by active transport (see Chapter 7). This allows the antibiotic to reach higher concentrations in the cell than in the medium and adds considerably to its antibacterial efficacy.

Tunicamycin n = 8, 9, 10 or 11

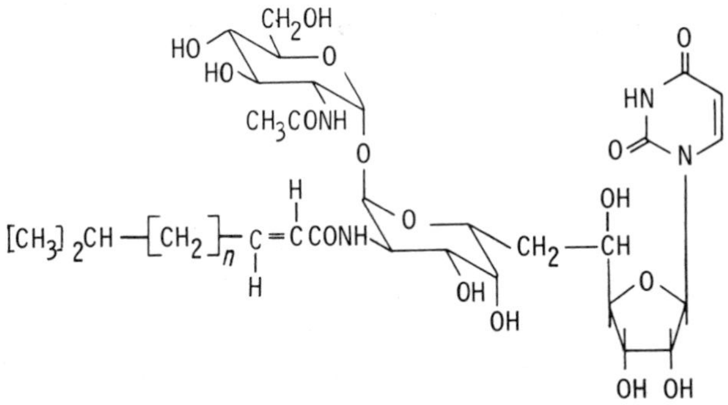

Figure 2.11 Tunicamycin.

Tunicamycin

Tunicamycin (Fig. 2.11) acts as an inhibitor of cell wall biosynthesis by specifically preventing the linkage between the undecaprenyl phosphate and compound V (Fig. 2.6). It also inhibits teichoic acid biosynthesis and the polymerization of carbohydrates in processes involving a long-chain acyl phosphate. The toxicity of this compound may be related to the fact that these latter reactions occur in animals and plants as well as bacteria.

Enduracidin, vancomycin and moenomycin

These three compounds are conveniently considered together because they affect that part of the peptidoglycan biosynthesis which occurs while the intermediates are bound to the cytoplasmic membrane, though each prob- ably affects different parts of this complex process. Each compound is chosen as a representative of a chemically related group of antibiotics. All are substances of relatively high molecular weight and act mainly against Gram-positive bacteria. With the exception of vancomycin, toxic side effects have limited the clinical usefulness of these agents.

Enduracidin A (Fig. 2.12) is a polypeptide antibiotic containing a number of unusual amino acid residues. It is cyclized through an ester bond and has a side chain terminated by an unsaturated acid group. Its name reflects its persistence in the animal body; a single dose maintains effective blood levels for more than 24 h. With peptidoglycan-synthesizing systems, either com- plete cells or as cell-free preparations to which UDP-N-acetylmuramyl pentapeptide and UDP-N-acetylglucosamine have been added as preformed substrates, addition of enduracidin leads to accumulation of the lipid in- termediate VIII (Fig. 2.6). The block in biosynthesis thus appears to occur beyond this point.

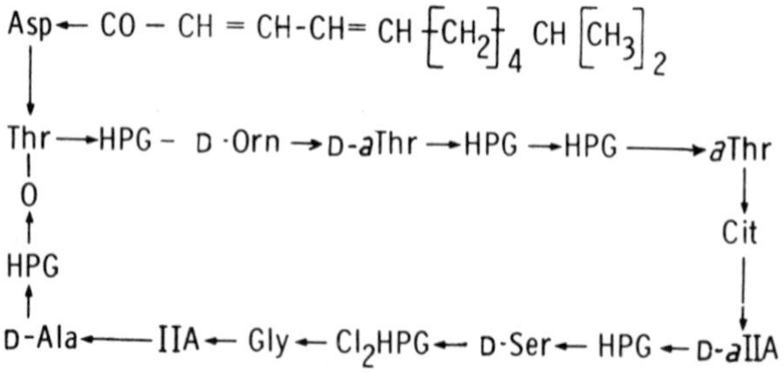

Asp ← CO – CH = CH-CH= CH $\left[CH_2\right]_4$ CH $\left[CH_3\right]_2$

Thr → HPG – D ·Orn → D-aThr → HPG → HPG ⟶ aThr
|
O
|
↑
HPG
|
↑
D-Ala ← ⟶ IIA ← Gly ← Cl$_2$HPG ← D·Ser ← HPG ← D-aIIA

(Cit below aThr, connected downward)

Figure 2.12 Enduracidin A. Abbreviations: HPG, *p*-hydroxyphenylglycine; Cl$_2$HPG, 3,5-dichloro-4-hydroxyphenylglycine; Cit, citrulline; Orn, ornithine; IIA, iminoimidazolinylaminopropionic acid:

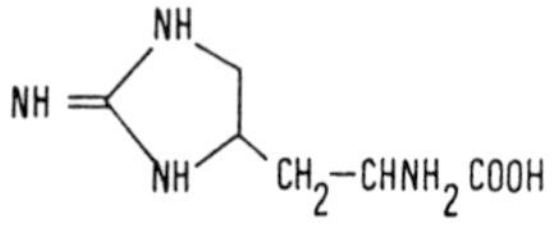

NH = (ring structure) NH ... NH CH$_2$–CHNH$_2$ COOH

a indicates the *allo* configuration. Amino acid residues are L unless otherwise indicated.

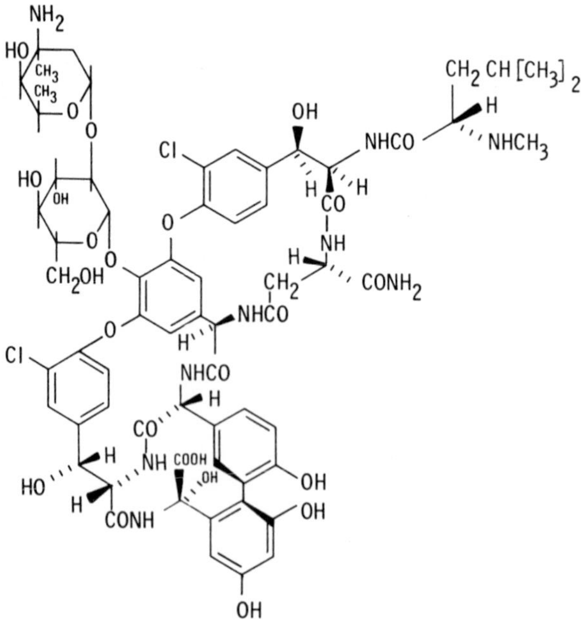

Figure 2.13 Vancomycin.

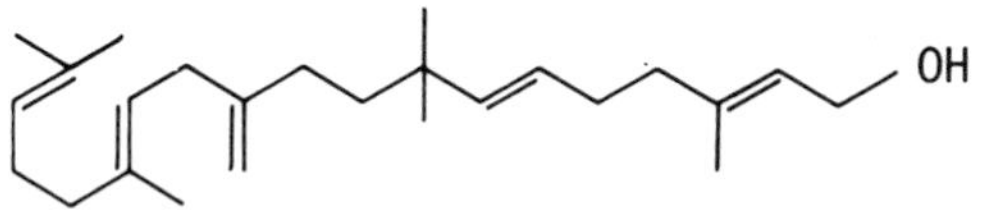

Figure 2.14 Moenocinol, a degradation product from moenomycin.

Of the group of antibiotics under discussion, vancomycin (Fig. 2.13) stands out as the only clinically important agent. This compound has been known for many years but its clinical importance has only emerged in the past 4–5 years, with the growing problem of methicillin-resistant staphylococci (MRS). The use of vancomycin is steadily increasing because of its value in this therapeutic area.

Vancomycin is also useful against troublesome intestinal infections due to *Clostridium difficile*. This organism sometimes multiplies and produces toxins when the usual gut flora have been largely eliminated by the use of broad-spectrum antibiotics. Vancomycin and the related ristocetin produce similar effects to enduracidin on peptidoglycan-synthesizing systems. The action of vancomycin depends on its ability to bind specifically to the D-alanyl-D-alanine group on the peptide side chain of the membrane-bound intermediates in peptidoglycan synthesis (compounds VI–IX in Fig. 2.6).

Moenomycin is representative of another large group of long-acting antibiotics having molecular weights of about 1800 and possessing one phosphorus atom per molecule. Various hydrolytic products including aminohexoses have been identified but the full structure is still unknown. One fragment is moenocinol (Fig. 2.14), a non-isoprenoid branched-chain unsaturated alcohol. This structure combined with the presence of a phosphate ester group has suggested some interaction with processes involving undecaprenyl phosphate. In peptidoglycan-synthesizing systems moenomycin resembles enduracidin in causing an accumulation of lipid intermediates.

It is thought that moenomycin inhibits one of the last steps in peptidoglycan biosynthesis, where the lipid-bound compound VIII (Fig. 2.6) is transferred to the nascent peptidoglycan.

Penicillins, cephalosporins and other β-lactam antibiotics

Penicillin was the first antibiotic to be discovered and used. It is one of a group of compounds known as the β-lactam antibiotics which are unrivalled in the treatment of bacterial infections. Their only serious defect is their characteristic of inducing immunological sensitization in a small proportion of patients, a side effect which precludes their use in those affected. The original penicillins isolated directly from mould fermentations were mixtures of compounds having different side chains. Addition of phenylacetic acid to

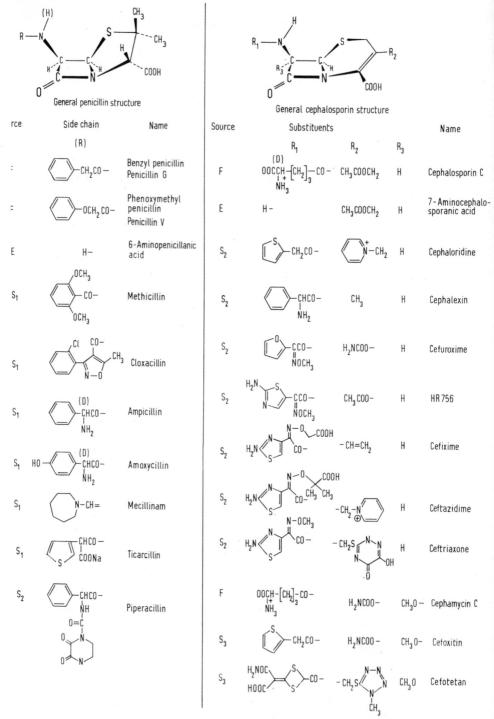

Figure 2.15 Penicillins and cephalosporins. Sources: F, by fermentation; E, by enzymic degradation of the corresponding fermentation product; S_1, by chemical synthesis from 6-aminopenicillanic acid; S_2, by chemical synthesis from 7-aminocephalosporanic acid; S_3, semi-synthetically from cephamycin C.

the fermentation medium improved the yield of penicillin and ensured that the product was substantially a single compound known as penicillin G or benzylpenicillin (Fig. 2.15). The first successful variant was obtained by replacing phenylacetic acid by phenoxyacetic acid as the added precursor. This gave phenoxymethylpenicillin or penicillin V (Fig. 2.15). The main advantage of this change was an improvement in the stability of the penicillin towards acid. The ready inactivation of penicillin G at low pH limited its usefulness when it was given by mouth, since a variable and often considerable fraction of the antibacterial activity was lost in the stomach. Penicillin V thus improved the reliability of oral dosing. These early penicillins, produced directly by fermentation, were intensely active against Gram-positive infections and gave excellent results in streptococcal and staphylococcal infections and in pneumonia. They were also very active against infections due to the Gram-negative gonococci and meningococci. These penicillins however are much less active against the more typical Gram-negative bacilli, and the concentration required for inhibiting their growth is about a thousand times as high as for Gram-positive organisms.

A further advance in the versatility of the penicillins was achieved by workers of the Beecham group with the development of a method for the chemical modification of the penicillin molecule. Bacterial enzymes were found that remove the benzyl side chain from penicillin G leaving 6-aminopenicillanic acid (Fig. 2.15) which could be isolated and then acylated by chemical means. This discovery opened the way to the production of an almost unlimited number of penicillin derivatives, some of which have shown important changes of properties compared with the parent penicillin. Three types of improvement have been achieved. The value of increased stability to acids has already been mentioned, and some semi-synthetic penicillins show this property. A limitation to the use of penicillins, particularly against staphylococci, was the rapid development of resistance due to the action of the enzyme penicillinase (β-lactamase) which converts penicillin to the antibacterially inactive penicilloic acid (see Chapter 8). Some modified penicillins (e.g. methicillin and cloxacillin, Fig. 2.15) are much less susceptible to attack by the commonest forms of this enzyme and are therefore useful against resistant strains.

The discovery of the β-lactamase inhibitor, clavulanic acid, (Fig. 2.16) provided an opportunity to co-administer this agent with β-lactamase-sensitive compounds. Examples of these mixtures are augmentin (a 1:1 mixture of amoxycillin and clavulanic acid) and timentin (a 1 : 1 mixture of ticarcillin and clavulanic acid). These preparations represent a significant step forward in penicillin therapy and help overcome some of the main limitations associated with these compounds.

The most striking change brought about by chemical modification of the penicillin side chain is an increase of activity against Gram-negative bac-

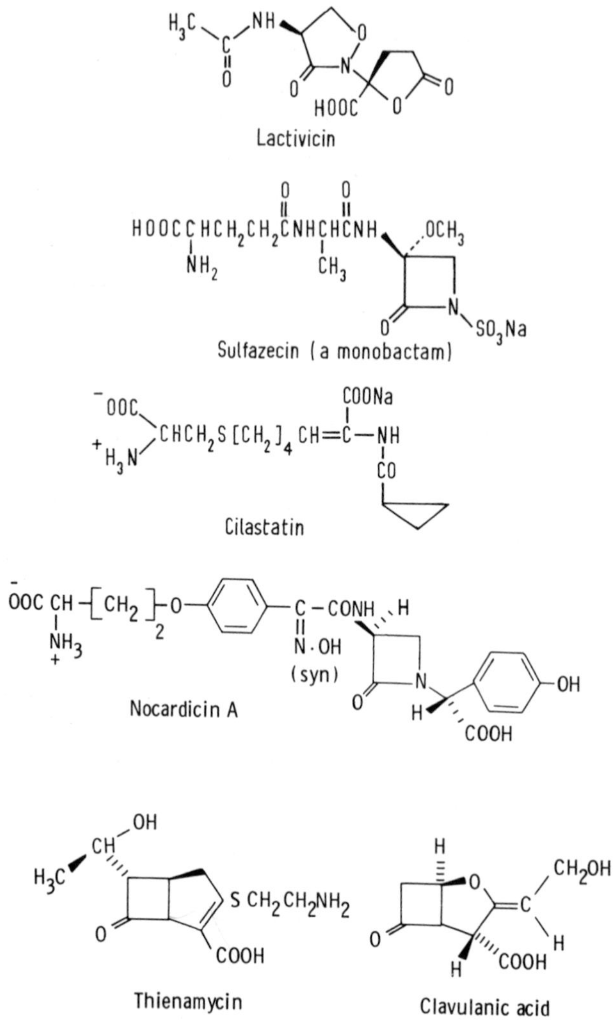

Figure **2.16** Some newer β-lactam antibiotics and an inhibitor of thienamycin metabolism (cilastatin).

teria, a property found in ampicillin, amoxycillin and piperacillin (Fig. 2.15). The increase in this type of activity is accompanied by a lessening of activity towards Gram-positive bacteria. The compounds thus approximate more closely to the broad-spectrum' type of antibiotics. Ampicillin is one of the most widely used antibacterial agents. In mecillinam (Fig. 2.15) the side chain is attached by an azomethine link rather than the usual amide bond. Here the activity spectrum of the original penicillin molecule has been completely reversed. This compound is highly active against Gram-negative bacteria but requires 50 times the concentration for an equal effect on Gram-positive organisms. It can be used in the treatment of typhoid fever.

Cephalosporin C (Fig. 2.15), isolated from a different organism from that used for penicillin production, was shown to have a structure similar in its nucleus to the penicillins. The biogenesis of the nuclei in these two classes of antibiotics is now known to be identical except that in cephalosporin biosynthesis the thiazolidine ring of the penicillin nucleus undergoes a specific ring expansion to form the dihydrothiazine ring of the cephalosporin nucleus. Besides this similarity in structure and biogenesis, cephalosporin C and its derivatives act on peptidoglycan cross-linking in the same way as the penicillins. Cephalosporin C itself has not found substantial use as an antibacterial drug. It is, however, amenable to modification in a similar way to the penicillins. Enzymic removal of the side chain gives 7-aminocephalosporanic acid (Fig. 2.15) which can be chemically acylated to give new derivatives. A second change in the molecule can also be made by a chemical modification of the acetoxy group of cephalosporin C. The first successful semi-synthetic cephalosporin was cephaloridine. Many others have followed, a selection of the best known being shown in Fig. 2.15. Most are only effective when given by injection, but cephalexin and cefixime can be given by mouth. Cefuroxime is unaffected by many of the common β-lactamases and can be used against bacterial strains which are resistant to other β-lactam antibiotics; it can be useful in infections due to *Neisseria* or *Haemophilus*. The related compound cefotaxime has enjoyed considerable success. Even newer agents such as ceftazidime and ceftriaxone (Fig. 2.15) are now finding favour because of the former's improved antipseudomonal activity and the latter's long half-life in the body which permits greater intervals between dosing.

The cephamycins resemble the cephalosporins, but have a methoxy group in place of hydrogen at position 7. Cefoxitin and cefotetan (Fig. 2.15) are semi-synthetic derivatives of cephamycin C. They are not readily attacked by β-lactamases and have advantages over the cephalosporin derivatives in activity against *Proteus* and *Serratia* species.

The enormous success of the penicillins and cephalosporins has stimulated a search for other β-lactam compounds. These have been found in a variety of micro-organisms. Some of the most interesting are shown in Fig. 2.16. In the carbapenem thienamycin the sulphur atom is not part of the ring, but is found in the side chain. This compound is remarkable for its high activity and broad antibacterial spectrum, but is inconveniently unstable. The N-formimidoyl derivative of thienamycin is significantly more stable and is used clinically. It is administered as a 1:1 mixture with cilastatin (Fig. 2.16). This latter compound is a potent inhibitor of renal dehydropeptidase (DHP) which would otherwise break down the carbapenem. Because of this potency and spectrum carbapenems are the subject of intense development activity in the pharmaceutical industry.

Other new β-lactams include the monobactams (Fig. 2.16) named as monocyclic bacterial β-lactams) which are derived from bacteria and represent the simplest β-lactam structures so far discovered to exhibit antibacterial activity. Many semi-synthetic derivatives have been made and some are in clinical trial. They exhibit excellent anti-Gram-negative activity with much weaker activity against Gram-positive bacteria. In contrast, the monocyclic nocardicins appear to offer less activity and are more of historic, rather than clinical, interest.

The most surprising recent discovery in this field has been lactivicin (Fig. 2.16). This γ-lactam structure appears in many test systems to behave as a β-lactam. It offers chemists a means of further understanding the complex relationship between structure and function in this class of antibacterial agents.

SITE OF ACTION OF PENICILLINS AND CEPHALOSPORINS

As with many other antibiotics, early attempts to discover the biochemical action of penicillin led to conflicting hypotheses. Gradually it became accepted that the primary site of action lay in the production of cell wall material, and more specifically in the biosynthesis of peptidoglycan. Even so, the determination of its exact point of attack proved difficult. Strominger and Tipper finally established the main features of the whole biosynthetic sequence, as already discussed, and provided evidence that penicillin acts on the transpeptidase which brings about the cross-linking (Fig. 2.7).

Evidence for this site of action rests on several different types of experiment. *Staphylococcus aureus* cells were pulse-labelled with [^{14}C]glycine, and peptidoglycan was isolated from their walls after growth for a further 20 min in unlabelled medium. The labelled glycine entered the pentaglycyl 'extending group'. The polysaccharide backbone of the peptidoglycan was broken down by an N-acetylmuramidase leaving the individual muramyl peptide units linked together only by their pentaglycyl peptide chains. The product was separated on columns of Sephadex G-50 and G-25. Radioactivity was distributed in a series of peaks of increasing molecular weight representing the distribution of the pulse of [^{14}C]glycine among peptide-linked oligomers of varying size. A parallel experiment done in the presence of penicillin showed the radioactivity to be associated very largely with a single low-molecular-weight peak, presumably the uncross-linked muramyl peptide unit, with much less activity in the oligomers. The penicillin had thus inhibited the peptide cross-linking.

In another experiment 'nucleotide pentapeptide' was prepared with D-[^{14}C]alanine. This was used as a substrate for a particulate enzyme preparation from *Escherichia coli* in the presence of UDP-*N*-acetylglucosamine. This system carried out the entire biosynthesis of peptidoglycan including the final cross-linking. Peptidoglycan was obtained as an insoluble product containing ^{14}C from the penultimate D-alanine of the substrate; the terminal D-[^{14}C]alanine was released into the medium, partly from the transpeptidase cross-linking reaction and partly from a carboxypeptidase that removed terminal D-alanine residues from cross-linked products. In a parallel experiment penicillin was added at a concentration that would inhibit growth of *Escherichia coli*. Biosynthesis of peptidoglycan then proceeded only to the stage of the linear polymer (IX, Fig. 2.6) which was isolated as a high-molecular-weight water-soluble product labelled with ^{14}C. No D-[^{14}C] alanine was liberated because the penicillin had suppressed both the cross-linking transpeptidase reaction and the action of DD-carboxypeptidase.

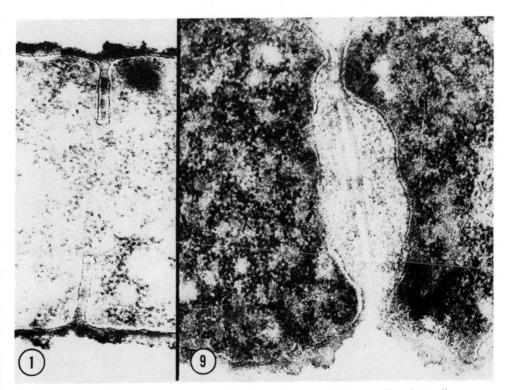

Figure 2.17 Electron micrographs of cross-sections of dividing cells of *Bacillus megaterium*. Picture (1) shows the septum forming in a normal cell. Picture (9) shows cell division after treatment with penicillin. Note the accumulation of fibrous material at the growing point. Reprinted by permission of The Rockefeller University Press from P. Fitzjames and R. Hancock (1965), *J. Cell Biol.*, **26**, 657.

Electron micrographs of thin sections of penicillin-treated *Bacillus megaterium* show abnormalities consistent with the accumulation of linear polymer in place of the normal cross-linked peptidoglycan. This is seen particularly during the formation of the septum which precedes the division of a cell. In Fig. 2.17 the appearance of a cross-section of normal septum formation is contrasted with a cross-section from a penicillin-treated cell. The normal wall is replaced by masses of fibrous material.

Release of uncross-linked peptidoglycan into the medium has been observed after penicillin treatment of growing cultures of *Micrococcus luteus* and a mutant of *Bacillus licheniformis*. In *Streptococcus pyogenes* pentapeptide subunits have been observed to accumulate near the septum during penicillin treatment.

The understanding of the mechanism of β-lactam action has been considerably advanced by the observation that there are groups of proteins to which β-lactams become covalently bound. These are referred to as penicillin-binding proteins or PBPs. In Gram-positive organisms there are four to five of these and in the Gram-negative enteric bacteria there are six main types. The properties of the *Escherichia coli* PBPs will be described in more detail and their characteristics are summarized in Table 2.1.

Different β-lactams exhibit different affinities for the various PBPs and these can, in turn, be correlated with different morphological effects as seen in Fig. 2.18. More detailed studies have recently revealed some fine structure in the PBP patterns on gel electrophoresis. It is now possible to see two major subgroups of PBP1, PBP1A and PBP1B. There are two proteins in the former groups and three in the latter. Each of these possesses both transpeptidase and transglycosylase activity. Compounds which bind most strongly to these PBPs cause cell lysis at the lowest antibacterial concen-

Table 2.1   Penicillin-binding proteins from *Escherichia coli*

Protein no.	$10^{-3} \times$ Molecular weight	Enzyme activity	Function
1A, 1B	91	Transpeptidase (wall) Transglycosylase	Cell elongation
2	66	Transpeptidase	Longitudinal growth
3	60	Transpeptidase (septum) Transglycosylase	Septum formation
4	49	Carboxypeptidase	Lipoprotein attachment
5	42	DD-carboxypeptidase	Limitation of
6	40	DD-carboxypeptidase	cross-linking

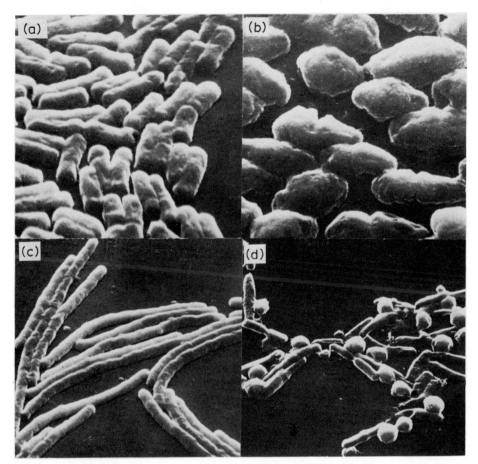

Figure 2.18 Effects of different β-lactam antibacterials on *Escherichia coli* shown by scanning electron microscopy. a, Normal culture. Other pictures show cells after 60 min treatment with the compounds indicated. b, Mecillinam (10 μg ml^{-1}). c, Cephalexin (32 μg ml^{-1}). d, Mecillinam (10 μg ml^{-1}) and cephalexin (32 μg ml^{-1}). Magnifications: a, 11 000; b, 12 000; c, 5800; d, 400. Reprinted by permission of the University of Chicago Press and the authors, D. Greenwood and F. O'Grady (1973), from *J. Infect. Dis.*, **128**, 793.

tration. PBP3 is involved in the synthesis of the septum peptidoglycan and also possesses transglycosylase activity. Compounds such as cephalexin, which bind more strongly to PBP3, inhibit the formation of septa and lead to the formation of filaments, which are greatly elongated cells (Fig. 2.18c). This distinction in the action of β-lactams associated with differences in binding behaviour adds further weight to the supposition that different mechanisms may be involved in synthesis of peptidoglycan for cell extension compared with its synthesis for septum formation. Yet another variation is found with mecillinam which binds almost exclusively to PBP2, and causes cells to assume an abnormal ovoid shape (Fig. 2.18b). PBP2 has proved more difficult to associate directly with a specific enzyme activity. PBP2-overproducing cells have been shown to possess enhanced amounts of cross-linked peptidoglycan and are very sensitive to mecillinam. It thus seems certain that PBP2 is responsible for the transpeptidase activity associated with the synthesis of this material. It only remains to be seen if there is any transglycosylase activity also associated with it, as is the case of the PBP1 and PBP3 transpeptidases discussed above. The most widely studied model β-lactam target enzymes have been the extracellular DD-carboxypeptidases produced by *Streptomyces* species and carboxypeptidases solubilized from the membrane of *Escherichia coli* or *Bacillus stearothermophilus*. The enzymes from *Streptomyces* display some transpeptidase activity besides their high carboxypeptidase activity. The relationship between these two enzymic actions has been shown with model substrates. The transpeptidase reaction which is exactly analogous to cross-linking of peptidoglycan (Fig. 2.7) is competitive with the carboxypeptidase reaction and is favoured by high concentrations of the substrates.

$$
\begin{array}{c}
\text{Transpeptidase} \\
\text{Ac}_2\text{L-Lys}\cdot\text{D-Ala}\cdot\text{D-Ala} + \text{Gly}\cdot\text{Gly} \xrightarrow{\quad\text{reaction}\quad} \text{Ac}_2\text{L-Lys}\cdot\text{D-Ala}\cdot\text{Gly}\cdot\text{Gly}
\end{array}
$$

Donor $\qquad\qquad\qquad\qquad\qquad\qquad$ + D-Ala

$$
+ \text{H}_2\text{O} \xrightarrow[\text{reaction}]{\text{Carboxypeptidase}} \text{Ac}_2\text{L-Lys}\cdot\text{D-Ala}
$$

Acceptors $\qquad\qquad\qquad\qquad$ + D-Ala

$(\text{Ac}_2\text{L-Lys} = N,N'\text{-diacetyl-L-lysine})$

Carboxypeptidases have been used to show the nature of the interaction of a penicillin or cephalosporin (I) with the enzyme (E). This may be represented as:

$$
\text{E} + \text{I} \underset{k_2}{\overset{k_1}{\rightleftharpoons}} \text{EI} \xrightarrow{k_3} \text{EI}^* \xrightarrow{k_4} \text{E} + \text{degraded inhibitor}
$$

The first step is reversible binding to the enzyme. The second stage, involving chemical modification of the inhibitor with covalent binding to the enzyme, is irreversible as is the final stage of enzyme release. For high antibacterial activity k_3 should be rapid, preventing release of I through reversal of the initial binding, and k_4 should be slow to maintain the enzyme in the inactive EI* form and to avoid significant re-activation. Measurements show that the widely used β-lactam antibiotics show just such characteristics, and this scheme goes far to explain their outstanding effectiveness. There is every reason to suppose that the inactivation mechanism is the same with cross-linking transpeptidases as with DD-carboxypeptidases. The nature of the end products of penicillin degradation depends on the enzyme involved. It may be a simple opening of the β-lactam ring to give the penicilloate (Fig. 8.4) or there may be more extensive breakdown leading to the production, from benzylpenicillin, of phenylacetylglycine. Those enzymes which yield penicilloate are equivalent to slow-acting β-lactamases. Quite possibly active β-lactamases are relatives of carboxypeptidases and transpeptidases in which reaction k_4 is rapid instead of very slow.

It seemed likely that the mechanism of action of DD-carboxypeptidases and cross-linking transpeptidases resembled that of certain esterases and amidases. These enzymes possess specially reactive groups, associated with their active centres, which undergo transient acylation in the course of enzymic action. Antibiotics containing a β-lactam ring behave chemically as acylating agents, reacting to produce penicilloyl derivatives. The action of penicillin on these enzymes might thus involve acylation of the enzymically active site with formation of the rather stable inactive complex EI*. This explanation has received strong support from experiments with purified DD-carboxypeptidases from *Bacillus stearothermophilus* and *Bacillus subtilis*. The enzyme was allowed to react briefly with [^{14}C]benzylpenicillin or with a substrate analogue [^{14}C]Ac$_2$L-Lys·D-Ala·D-lactate; D-lactic acid is the exact hydroxyl analogue of D-alanine, and use of this derivative enables the transient enzyme reaction intermediate to be trapped. The treated enzyme was then broken down into peptide units either by cyanogen bromide or pepsin. In products from the *Bacillus stearothermophilus* enzyme, radioactivity was found in a peptide with 40 amino acid residues. This was sequenced, and the label was shown to be associated with the same specific serine residue, whether the reactant was benzylpenicillin or the substrate analogue. Similar results were found with the *Bacillus subtilis* enzyme from which a labelled 14-unit peptide was isolated. This peptide showed strong homology with 14 residues of the *Bacillus stearothermophilus* peptide and the label was associated with the corresponding serine residue. Thus for these enzymes penicillin apparently binds to the active site and acylates the same serine as the substrate. Unlike the substrate, the degraded penicillin is only released very slowly (reaction k_4 in the scheme above) and thus blocks further access of substrate to the site.

How can this action of penicillin be related to its structure? The most widely quoted explanation is due to Strominger. It depends on the similarity of the spatial orientation of the principal atoms and polar groups in the penicillin nucleus to one particular orientation of the D-alanyl-D-alanine end group of the pentapeptide side chain of peptidoglycan precursors (see Fig.

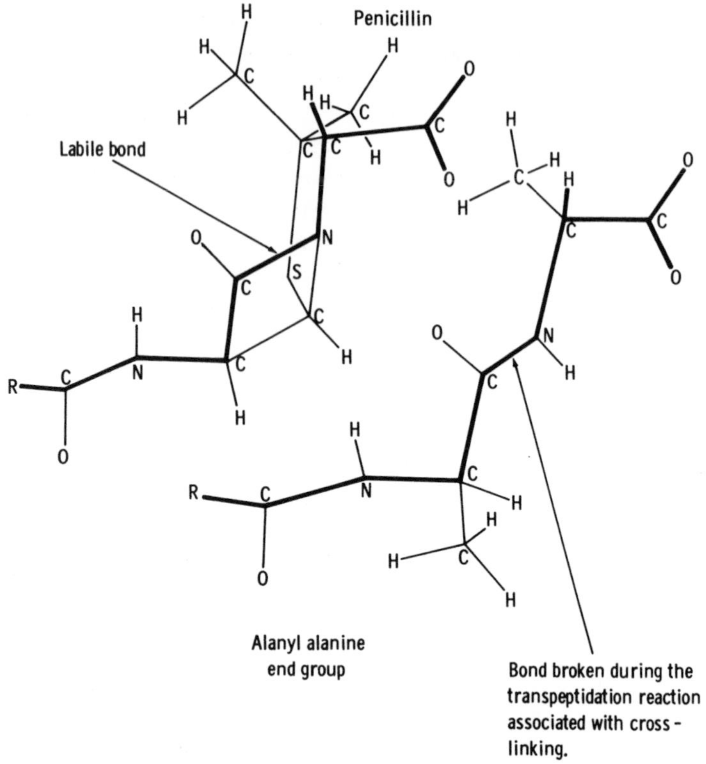

Figure 2.19 Comparison of the structure of penicillin with that of the D-alanyl-D-alanine end group of the peptidoglycan precursor. Reproduced by permission of the Federation of American Societies for Experimental Biology from J. L. Strominger *et al.* (1967) *Fed. Proc.*, **26**, 18.

2.19). When the two structures are compared, the peptide bond between the alanine units is seen to correspond in position to the lactam group in the four-membered ring of penicillin responsible for its acylating properties. Such a group bound to the cross-linking transpeptidase close to its active centre could well usurp the acylating function implicit in the normal reaction of the substrate with the enzyme. When the structures (illustrated in Fig. 2.19) are compared more critically it becomes apparent that the agreement between them is imperfect, but can be much improved if the peptide bond of the D-alanyl-D-alanine end group is represented not in its normal planar form but twisted nearly 45° out of plane. This may imply that the conformation of the penicillin molecule resembles the transition state of the substrate rather than its resting form. During the enzymic transpeptidation the peptide bond quite possibly undergoes this sort of distortion. A further discrepancy is that the usual penicillins lack a methyl group in a position corresponding to that of the penultimate D-alanine residue. A penicillin with a methyl group in the appropriate 6-position, which would be expected to give a better fit, was made and shown to be antibacterially inactive, though oddly the 6-methoxy derivatives show good activity. Clearly there is still much to be learnt about the exact relationship between structure and mode of action. The rigidity of the bicyclic ring structures of the penicillins and cephalosporins maintains the principal binding groups in fixed relative conformation. This was thought to be an important feature in binding to the active site. However, even this concept may have to be reviewed in light of the discovery of lactivicin and the monobactams.

FURTHER READING

Chase, H. A., Fuller, C. and Reynolds, P. E. (1981) The role of PBPs in the action of cephalosporins against *Escherichia coli* and *Salmonella typhimurium*. *Eur. J. Biochem*, **117**, 301.

Frere, J-M. and Joris, B. (1985) Penicillin sensitive enzymes in peptidoglycan biosynthesis. *CRC Crit. Rev. Microbiol.*, **11**, 299.

Gallego, A. and Rubio, J. M. (eds) (1977) Fosfomycin (a symposium on phosphomycin). *Chemotherapy*, **23**, Suppl. 1.

Greenwood, D. (1986) An overview of the response of bacteria to β-lactam antibiotics. *Rev. Infect. Dis.*, **8**, Suppl. 5, S487.

Gutmann, L., Vincent, S., Billot-Klein, D., Acar, J. F., Mrina, E. and Williamson, R. (1986) Involvement of PBP2 with other PBPs in lysis of *Escherichia coli* by some β-lactam antibiotics alone and in synergistic lytic effect of Amdinocillin (mecillinam). *Antimicrob. Ag. Chemother.*, **30**, 906.

Hackenbeck, R., Holtje, J-V. and Labischinski, H. (eds) (1983) *The Target of Penicillin – The Murein Sacculus of Bacterial Cell Walls – Architecture and Growth*, Walter de Gruyter.

Lowe, G. and Swain, S. (1985) Do β-lactams require a β-lactam ring? *Spec. Publ. R. Soc. Chem.*, **52**, 209.

Nakagawa, J., Tamaki, S., Tomioka, S. and Matsuhashi, M. (1984) Functional biosynthesis of cell wall peptidoglycan by polymorphic bifunctional polypeptides. *J. Biol. Chem.*, **259**, 13937.

Osborn, M. J. and Wu, H. C. P. (1980) Proteins of the outer membrane of Gram-negative bacteria. *Annu. Rev. Microbiol.*, **34**, 369.

Seidl, P. H., Zwerenz, P., Golecki, J. R. and Schleifer, K. H. (1985) *Streptococcus pyogenes* grown under sub-lethal concentrations of penicillin G accumulates pentapeptide subunits of peptidoglycan close to the septum. *FEMS Microbiol. Lett.*, **30**, 325.

Sutherland, I. W. (1985) Biosynthesis and composition of Gram-negative bacterial extracellular and wall polysaccharides. *Annu. Rev. Microbiol.*, **39**, 243.

Ward, J. B. (1977) Tunicamycin inhibition of bacterial wall polymer synthesis. *FEBS Lett.*, **78**, 151.

Wyke, A. (1984) Isolation of five PBPs from *Staphylococcus aureus*. *FEMS Microbiol. Lett.*, **22**, 133.

3 Antiseptics, antibiotics and the cell membrane

MICROBE KILLERS: ANTISEPTICS AND DISINFECTANTS

The major interest throughout this book lies in the mechanism of action of drugs that can be used against microbial infections. For this purpose the compound must normally be absorbed and circulate in the blood. However, there is also a requirement in medicine and in industry for substances that kill bacteria and other micro-organisms on the surface of the body or in other places. Such products are known as disinfectants, sterilants, antiseptics or biocides, the choice of term depending on the circumstances in which they are used. 'Disinfectant' describes products intended for use in the presence of dirt and dense bacterial populations, for example in the cleaning of animal quarters or drains. 'Biocide' is used more particularly for preservatives that prevent bacterial and fungal attack on wood, paper, textiles and all kinds of organic material. 'Antiseptic' is a term usually reserved for a substance that can be safely applied to the skin and mucosal surfaces with the aim of reducing the chances of infection by killing the surface bacteria. 'Sterilants' are substances used to sterilize an enclosed space; since penetration is paramount in this application sterilants are usually gaseous. However, all these terms are to some extent overlapping and interchangeable.

The requirements for a compound having disinfectant or antiseptic action differ markedly from those needed in a systemic drug. Many compounds sucessfully used against bacterial infections do not actually kill bacteria, but only prevent their multiplication; most are inactive against non-growing bacteria. A cessation of bacterial growth is usually all that is needed in treating an infection, since the body has antibody and phagocytic defences that can soon be mobilized to remove bacteria present in relatively small numbers. Furthermore, systemic antibacterials are often active against a fairly limited range of bacteria. This is acceptable since the compound can be selected according to the nature of the infection that is being treated. Antiseptics, by contrast, are usually required to have a broad-spectrum, killing effect.

A distinction is often made between bacteriostatic and bactericidal compounds, but the division is by no means clear cut. There is no certain way of determining whether a bacterium is dead. The usual method of assessing the killing effect of an antiseptic is by measuring the 'viable count' of a treated bacterial suspension. The antiseptic is first inactivated and dilutions of the suspension are added to a rich medium. Bacteria are deemed to be alive if they give rise to colonies. Many compounds are bacteriostatic at low concentrations and bactericidal at higher concentrations, and the effect may also depend on the conditions of culture. However, for antiseptics and disinfectants a bactericidal effect is required under all normal conditions of application. Such compounds must be able to kill bacteria whether they are growing or resting and must be able to deal with most of the common bacteria likely to be found in the environment and, ideally, fungi and viruses as well. Bacterial and fungal spores are usually much more difficult to kill.

Many of the older disinfectants are compounds of considerable chemical reactivity. Their antimicrobial action presumably depends on their ability to react chemically with various groups on or in the organism thus killing it. Such compounds include hydrogen peroxide, the halogens and the hypochlorites, the gaseous sterilants ethylene oxide, ozone, etc. Salts ands other derivatives of the heavy metals, particularly of mercury, probably owe their antimicrobial effect to reaction with vital thiol groups. Although some of these substances still have important uses in disinfection, their high reactivity and toxicity limit their scope and they are not nowadays generally acceptable for the more delicate uses as antiseptics. For this purpose two main groups of compounds are used almost exclusively, the phenols and the cationic antiseptics. The main emphasis with these agents has been their efficacy against bacteria. Increasingly, however, there is concern that they should have useful activity against fungi and viruses. Although there are differences between the actions of phenols and cationic antiseptics they show many common features. The literature on the mode of action of antiseptics is confused, and much of the earlier work needs revising in the light of more modern techniques. A review of the most reliable evidence suggests the following generalizations.

(1) Antiseptics bind readily to bacteria, the amount absorbed increasing with an increasing concentration in solution. The adsorption isotherm sometimes shows a point of inflection which corresponds to the minimum bactericidal concentration; higher concentrations lead to a much greater adsorption of the compound. The most important site of adsorption is the cytoplasmic membrane. Spheroplasts or protoplasts lacking the outer cell wall layers will bind the antiseptic and may be lysed or damaged. Adsorption by isolated cell membranes can also be demonstrated.

(2) The extent of killing of the bacteria is governed by three principal factors: (a) concentration of the antiseptic, (b) bacterial cell density, (c) time of contact. The adsorption of a given amount of the compound per cell leads

to the killing of a definite fraction of the bacterial population in a chosen time interval.

(3) The lowest concentration of the antiseptic that causes death of the bacteria also brings about leakage of cytoplasmic constituents of low molecular weight. The most immediately observed effect is loss of potassium ions. Leakage of nucleotides is often detected by the appearance in the medium of material having an optical absorption maximum at 260 nm. Gram-positive cells show leakage of amino acids. Some loss of cytoplasmic solutes is not in itself lethal. Compounds are known that cause this effect but do not kill bacteria; moreover bacteria that have been rendered leaky by low concentrations of an antiseptic will often grow normally if they are immediately washed and placed in a nutrient medium. The increased permeability is a sign of changes in the membrane which are initially reversible but become irreversible on prolonged treatment.

(4) The necessary characteristic of antiseptics is their bactericidal action, but there is often a low and rather narrow concentration range in which their effect is bacteriostatic. At these low concentrations certain biochemical functions associated with the bacterial membrane may be inhibited.

(5) In the presence of higher concentrations of antiseptic and after prolonged treatment, the compound usually penetrates the cell and brings about extensive ill-defined disruption of normal cellular functions.

The primary effect of these antiseptics on the cytoplasmic membrane is thus established beyond doubt, but secondary actions on cytoplasmic processes are less defined and may vary from one compound to another. Examples of evidence of action for particular compounds will be given as illustrations.

Phenols

Crude mixtures of cresols solubilized by soap or alkali and originally introduced as 'lysol' are still used as rough disinfectants. They need to be applied at high concentrations and are irritant and toxic but they kill bacteria, fungi and some viruses. For more refined applications as antiseptics chlorinated cresols or xylenols are commonly used. These compounds are less active than the cationic antiseptics against staphylococci and pseudomonads. Hexachlorophene (Fig. 3.1) is a different kind of phenolic antiseptic. It acts only slowly, but binds strongly to the skin. It was used widely in surgical soaps and antiperspirant preparations. However, absorption through the skin can cause damage to the central nervous system, particularly in infants, and use of the compound is now severely restricted.

Hexachlorophene at bactericidal concentrations causes release of cytoplasmic components absorbing at 260 nm. A change in permeability to compounds external to the cell can be demonstrated using N-tolyl-1-naphthylamine-8-sulphonic acid (tolylperi acid), a compound that fluoresces

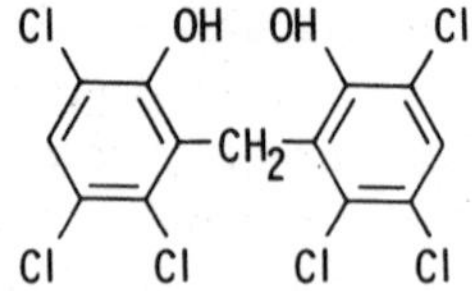

Hexachlorophene

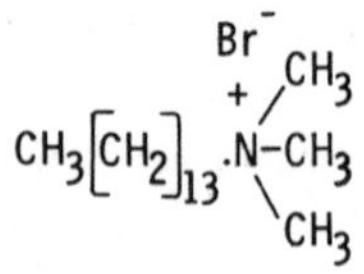

Cetrimide

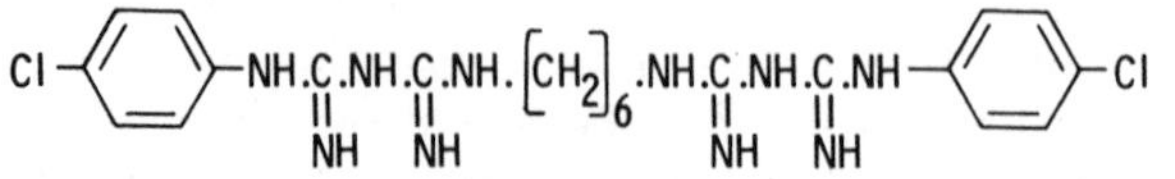

Chlorhexidine

Figure 3.1 Synthetic antiseptics. The formula for cetrimide shows the main component in the preparations normally sold. Homologues with other chain lengths, especially C_{16}, are also present.

strongly when it binds to proteins. This compound does not penetrate into normal bacteria, but bacteria treated with hexachlorophene absorb it and show a brilliant fluorescence. The uptake of 3,5,3′,4′-tetra-chlorosalicylanilide by *Bacillus megaterium* has been studied with the aid of ^{14}C-labelled material. Cell membranes were isolated and freed from other cellular components; they were shown to carry the whole of the radioactivity bound by the cells. The same compound tested at sublethal concentrations on *Staphylococcus aureus* caused a marked inhibition of oxygen consumption. This was interpreted as evidence of a direct effect on the respiratory mechanism which is associated with the membrane. Inhibition of other biochemical functions, e.g. fermentation, required higher concentrations which caused cytoplasmic leakage and cell death.

Alcohols

Alcohols are still used as cheap disinfectants and preservatives. Ethanol, for example, is a reasonably effective skin disinfectant as a 60–70% solution. Isopropanol (propan-2-ol) (50–60%) can be used to sterilize instruments such as clinical thermometers and the more complex compound known as bronopol (2-bromo-2-nitropropane-1,3-diol) is an effective preservative for certain pharmaceutical products and toiletries.

The antibacterial effects of the alcohols can be traced to a disruption of membrane function. The action of short-chain alcohols such as ethanol is probably dominated by the polar function of the hydroxyl group which may form a hydrogen bond with the ester groups of membrane fatty acid residues. In contrast with ethanol, longer-chain alcohols gain access to the hydropho-

bic regions of membranes and this probably accounts for the increasing
potency of antimicrobial action up to a maximum chain length of 10 carbon
atoms. The interaction of alcohols with cell membranes produces a generalized increase in permeability which is lethal at higher concentrations.
Bronopol may exert an additional effect by interacting with thiol groups in
membrane proteins.

Cationic antiseptics

This classification covers a number of compounds differing considerably in
chemical type. Their common features are the presence of strongly basic
groups attached to a fairly massive lipophilic molecule. Although antiseptic
action is found quite widely in compounds having these characteristics the
degree of activity is sharply dependent on structure within any particular
group. For instance in cetrimide (Fig. 3.1) the length of the main alkyl chain
is 14 carbon atoms and the activity of other compounds in the same series
falls off markedly with longer or shorter chains. Cetrimide combines excellent detergent properties with a useful antiseptic action, although it is not
very potent against *Proteus* and *Pseudomonas* species and has little antiviral
activity, except against viruses with a lipid envelope. Experiments with
Escherichia coli labelled with ^{32}P have shown that with increasing concentrations of cetrimide the loss of cell viability closely parallels the degree of
leakage of radioactivity from the bacteria. An effect on bacterial growth
however is noticeable at concentrations that affect neither viability nor
permeability.

One of the best and most widely used of the cationic antiseptics is
chlorhexidine (Fig. 3.1). This compound has two strongly basic groups, both
biguanides; it is often formulated as the digluconate which has good solubility in water. Chlorhexidine is much less surface active than cetrimide and has
little detergent action. However, it acts against a wide range of bacteria at
concentrations between 10 and 50 μg ml^{-1} and it also has useful activity
against *Candida albicans*. Its toxicity is low and it has so little irritancy that it
can be used on the most sensitive mucosal surfaces. For example, it is a useful
aid to oral hygiene. Periodic rinsing of the mouth with chlorhexidine solution
greatly reduces the population of *Streptococcus mutans* on the teeth. This
minimizes the production of dental plaque and reduces periodontal infections that give rise to gingivitis. It also decreases the incidence of some types
of caries. An important feature of this action is the strong binding of
chlorhexidine to the tissues in the mouth including the teeth with subsequent
slow release which maintains an antibacterial action over an extended
period.

Chlorhexidine shows effects on the cytoplasmic membrane characteristic
of cationic antiseptics. At concentrations that just prevent growth of *Strepto-*

coccus faecalis it inhibits the adenosine triphosphatase of the membrane. The effect can be shown in isolated membranes or on the solubilized enzyme derived from it. A similar concentration of chlorhexidine inhibits the net uptake of potassium ions by the intact cells, and the two effects are thought to be interdependent. This interaction beween chlorhexidine and the cell membrane probably involves electrostatic binding between the cationic

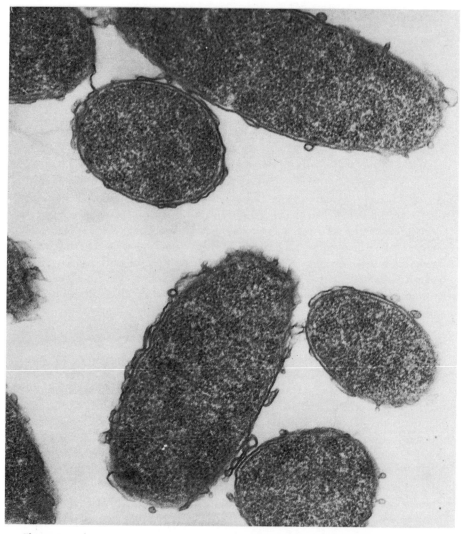

Figure 3.2 Electron micrograph of a cross-section of an *Escherichia coli* cell after treatment with a low concentration (30 μg ml^{-1}) of chlorhexidine, showing 'blistering' of the cell wall. We thank Mr A. Davies and Mrs M. Bentley for this picture.

groups of the antiseptic and anionic residues in the membrane. Hydrophobic interactions between the hexamethylene chains of chlorhexidine and the aliphatic chains of membrane lipids also contribute to the stability of the complex. When bacteria are treated with a range of concentrations of chlorhexidine and then examined for leakage of cytoplasmic solutes, the degree of leakage increases with concentration up to a maximum and then declines at higher concentrations. Low concentrations of chlorhexidine provoke the release of K^+ ions, nucleotides and sugars. Electron microscopy shows that the cells from higher levels of chlorhexidine treatment are grossly altered. The increased membrane permeability apparently allows the antiseptic to enter the cytoplasm and to cause precipitation of the nucleic acids and proteins resulting in the death of the cells. Under these circumstances leakage is probably prevented by simple mechanical blockage.

With Gram-negative bacteria chlorhexidine may affect the outer membrane as well as the cytoplasmic membrane. This shows itself in a 'blistering' appearance which can be seen in electron micrographs (Fig. 3.2). This phenomenon will be discussed further in connection with the action of polymyxin.

POLYPEPTIDE ANTIBIOTICS

Several classes of polypeptide antibiotics are known. In two groups the effects on bacteria conform exactly with the properties already discussed for the phenolic and cationic antiseptics. They owe their primary antibacterial action to their binding to the cytoplasmic membrane, with subsequent disturbance of its function and can therefore be regarded as cytolytic agents. Both types are cyclic polypeptides. One group includes the tyrocidins and gramicidin S which are cyclic decapeptides (Fig. 3.3). These contain one or sometimes two free amino groups. They are more active against Gram-positive than against Gram-negative bacteria. The polymyxins which form the second group have a smaller polypeptide ring attached to a polypeptide chain terminating with a branched 8- or 9-carbon fatty acid residue. They have five free amino groups associated with the diaminobutyric acid units. The octapeptins have a similar ring structure but differ in the composition of the side chain. The antibacterial action of polymyxins and octapeptins is directed particularly against Gram-negative organisms. This selectivity can be dramatically altered by chemical modification. Thus, the penta-N-benzyl derivative is highly active against Gram-positive bacteria.

These polypeptide antibiotics have only a minor place in medicine because they also damage mammalian cell membranes. The polymyxins may be used systemically in severe *Pseudomonas* infections, though there is considerable risk of kidney damage.

Polymyxin is bactericidal and acts on non-growing as well as on growing

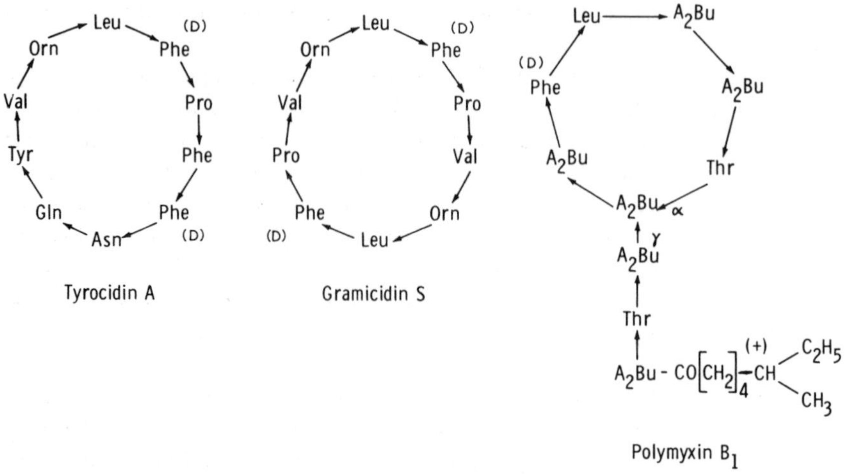

Figure 3.3 Antibiotics that produce effects on bacterial cell membranes. The usual abbreviations are used for the common amino acids. A_2Bu, 2,4-diaminobutyric acid. Arrows show direction of the peptide bond —CO—NH—. Except where shown, all peptide linkages involve α-amino and α-carboxyl groups. Configurations are L unless otherwise indicated.

cells. At low concentrations its bactericidal action parallels the degree of release of cytoplasmic solutes. It is strongly and rapidly bound to bacteria. With *Salmonella typhimurium* the binding of 2×10^5 molecules of polymyxin per cell was shown to be bactericidal. The site of this binding has been studied with the aid of a derivative in which a fluorescent adduct is attached to one of the free amino groups (up to two of these groups may be acylated without loss of antibacterial effect). When *Bacillus megaterium*, an organism sensitive to the antibiotic, is treated with this fluorescent derivative the cells become fluorescent. Fractionation of the cells shows the fluorescence to reside almost entirely in the cytoplasmic membrane. An increase in the permeability of polymyxin-treated *Pseudomonas aeruginosa* towards solutes in the medium has been shown in an experiment with tolylperi acid similar to that described for hexachlorophene (see above). With this same organism low concentrations of polymyxin inhibit respiration, an effect also observed with some phenolic antiseptics.

Physical measurements of various kinds all tend to the conclusion that the antibacterial action of polymyxin is caused primarily by its binding to the cytoplasmic membrane. The positively charged peptide ring is thought to bind electrostatically to the anionic phosphate head groups of the membrane phospholipid, displacing magnesium ions which normally contribute to membrane stability. At the same time the fatty acid side chain is inserted into the hydrophobic inner region of the membrane. The effect is to disturb the normal organization of the membrane and to alter its permeability character-istics.

In Gram-negative bacteria there is a further binding to the outer mem-brane affecting mainly the lipopolysaccharide but with similar disorganizing effects. The gross effects of polymyxin on the outer membrane are sometimes revealed in electron micrographs as blisters similar to those caused by chlorhexidine (Fig. 3.2). The swellings may be due to an increase in the surface area of the outer leaf of the outer membrane. The parallels between the action of polymyxin and chlorhexidine are quite striking. In both, the binding and antibacterial effects can be antagonized by excess of calcium or magnesium ions, indicating that the displacement of divalent ions is an important feature of their action.

The tyrocidins are also bactericidal and promote leakage of cytoplasmic solutes. Their action on the bacterial membrane permits passage into the cell of ions that are normally excluded, and under some conditions this causes uncoupling of oxidative phosphorylation as a secondary effect. Gramicidin S, a closely related compound, acts similarly. It lyses protoplasts from *Micrococcus lysodeikticus* but not those from *Bacillus brevis*. Since it is bactericidal towards the former organism but not the latter it is reasonable to suppose that both its action and specificity depend upon its effect on the cytoplasmic membrane. The tyrocidins act not only on bacteria but also on the fungus *Neurospora crassa*. In this organism concentrations of the antibiotic that stop growth and cause leakage of cell contents also cause an immediate fall in membrane potential, a consequence of the destruction of the permeability barrier.

In both the tyrocidin group and in the polymyxins the cyclic molecular structure is important for antibacterial activity. The presence of basic groups is also essential, but in other respects the molecules can be varied consider-ably without losing activity. The simple symmetrical structure of gramicidin S has been subjected to many modifications. Activity is preserved when the ornithine units are replaced by arginine or lysine groups but is lost by modifications destroying the basic character of the terminal groups. The compound in which glycine replaces L-proline is fully active. Moreover one L-proline residue together with the adjacent D-phenylalanine can be replaced by a δ-aminopentanoic acid group without losing antibacterial activity. The resulting compound has only nine peptide groups, but retains the same ring size. Acyclic compounds having the same sequence of amino acids as gramicidin S show only slight antibacterial action.

The importance of the cyclic structure lies in the maintenance of a well-defined compact conformation in solution. This has been shown by nuclear magnetic resonance, optical rotatory dispersion and other physical measurements. In tyrocidin A and gramicidin S the conformation is deter-mined by lipophilic association between the non-polar side chains of the

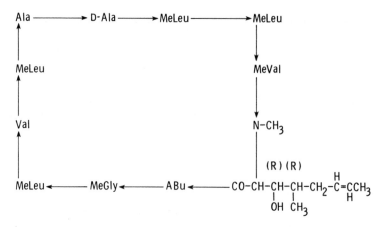

Figure 3.4 Cyclosporine (cyclosporin A).

amino acids, particularly leucine, valine, proline and phenylalanine, and by hydrogen-bonding between the peptide groups. Three regions have been defined in the molecular topography of tyrocidin A: a hydrophobic surface, a flat hydrophilic opposite surface consisting of the peptide groups of most of the amino acids in equatorial positions and a helical hydrophilic region accommodating the amide groups of asparagine and glutamine and the tyrosine hydroxyl group. Gramicidin S shows a similar arrangement, based on a pleated-sheet structure. In both antibiotics the ornithine amino groups, which are essential for antibacterial activity, stand out from the hydrophilic surface.

Cyclosporine formerly called cyclosporin A (Fig. 3.4) is an unusual cyclic polypeptide antibiotic in which seven of the 11 amino acid units are N-methylated. It was originally discovered as an antifungal agent, but has proved to have remarkable immunosuppressive properties and in conjunction with other immunosuppressive drugs, it is used successfully to prevent rejection in organ grafts. The drug blocks the activation of T-lymphocytes by an unknown mechanism. Efforts to identify a protein receptor for cyclosporine have been unsuccessful and the hydrophobic nature of the molecule suggests that it may interact with the lipid phase of the cell membrane to increase fluidity and impair the membrane potential changes involved in lymphocyte activation.

IONOPHORIC ANTIBIOTICS

Several classes of antibiotics may be grouped together because of their common property of facilitating the passage of inorganic cations across membranes by formation of hydrophobic complexes with the ions or by forming ion-permeable pores across the membranes. Although these compounds were discovered through their antibacterial activity, they are not used in medicine because of their lack of specificity. They act equally effectively on the membranes of animal cells and may therefore be toxic. Nevertheless they are of considerable biochemical interest and are widely used as experimental tools. As antibacterial agents they are active mainly against Gram-positive organisms. Gram-negative bacteria are relatively insensitive because their outer membranes are impermeable to hydrophobic compounds of the molecular size of the ionophores.

Valinomycin

This was the first member of a group of related compounds to be discovered and the most widely studied. It is a cyclic depsipeptide in which amino acids alternate with hydroxy acids in a ring which contains both peptide and ester groups (Fig. 3.5). An important feature is the alternation of D- and L-configurations in pairs around the 12 components of the ring structure. This is a common feature of all the cyclic ionophores; where the ring has six components, as in enniatin, the configuration alternates between each component. Valinomycin forms a well-defined complex with potassium ions. This crystalline compound has been studied by X-ray diffraction and shows a highly ordered structure (Fig. 3.6) in which the potassium atom is surrounded by six oxygen atoms. The ring structure is puckered and held in a cylindrical or bracelet-like form by hydrogen bonds roughly parallel to its axis. The ability to achieve such a conformation depends entirely on the

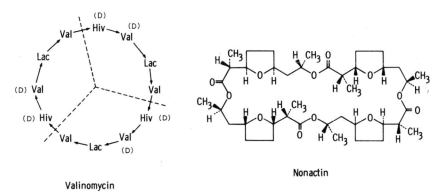

Valinomycin

Nonactin

Figure 3.5 Antibiotics affecting the permeability of membranes to potassium ions. In the valinomycin structure residues are represented thus: Val, valine; Lac, lactic acid; Hiv, 2-hydroxyisovaleric acid. Arrows show the direction of peptide or ester bonds, —CO—NH— or —CO—O—. The asymmetric centres have L configuration unless otherwise indicated. Dotted lines separate the repeating units.

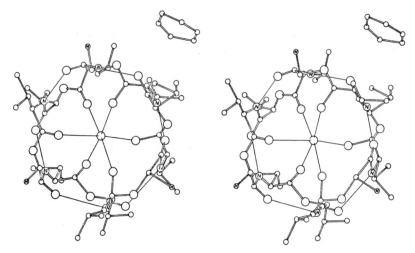

Figure 3.6 Stereophotographs of a model of the potassium complex of valinomycin. To obtain a three-dimensional effect the diagram should be held about 20 inches from the eyes and attention concentrated on the space between the two pictures. With practice three pictures can be seen, the middle one showing a full stereoscopic effect. The central metal ion is seen co-ordinated to six oxygen atoms. Nitrogen atoms are labelled N and the methyl groups of the lactyl residues M. Hydrogen bonds are shown by thin lines. The solitary hexagonal ring is hexane of crystallization. The diagram was generated from X-ray crystallographic data by a computer program written by Carroll K. Johnson, Oak Ridge, Tennessee. We express our gratitude to Mary Pinkerton and L. K. Steinrauf for supplying this illustration and allowing us to publish it.

alternation of D- and L-centres. The dimensions are such that the potassium atom is exactly accommodated. The ion entering the complex must shed its normal hydration shell; the complex retains the positive charge carried by the ion. The structure observed in the crystal is substantially maintained in solution. Although valinomycin will also form a complex with sodium, the smaller sodium atom fits much less exactly into the structure and this complex has a stability constant one thousand times smaller than that of the potassium complex.

The high specificity of valinomycin towards the potassium ion and the physical properties of the complex are in accord with its postulated action on biological membranes. The binding of the potassium ion in the structure of valinomycin increases the lipophilicity of the antibiotic and thereby promotes its diffusion into the hydrophobic regions of the membrane. Much of the work on valinomycin has been done with artificial membranes, liposomes and mitochondria. In all these the essential action of valinomycin is to facilitate the transport of potassium across the membrane. The lipophilic molecule moves physically through the membrane lipids carrying potassium and returns in the protonated form. In a passive membrane the flow is determined solely by the concentration of potassium ions on each side of the membrane, but in mitochondria supplied with an energy source potassium is taken in by an energy-coupled process against the concentration gradient. The process is highly effective, one valinomycin molecule being able to transport 10^4 ions s^{-1}, a turnover rate higher than that of many enzymes. Kinetic studies reveal that the transport of potassium by valinomycin and similar ionophores shows saturation kinetics with respect to the cation; sodium ions inhibit potassium transport although they undergo little transport themselves. The kinetic results are well explained by a model in which the ionophore at the membrane surface first forms a hydrophilic cation complex. This is transformed to a hydrophobic complex which can then cross the membrane. The rate of the transformation from one type of complex to the other determines the turnover number.

The mechanism of the inhibitory action of valinomycin on bacteria has received less study. In the case of *Streptococcus faecalis* growing anaerobically there is no oxidative phosphorylation mechanism; ATP is generated solely by glycolysis. Valinomycin is inhibitory to this organism in normal media of low potassium content. It specifically drains the cell of potassium and growth ceases because of the requirement for potassium in cellular metabolism. If the potassium content of the medium is raised to that normally present in the cytoplasm the inhibitory action of valinomycin is prevented. With growing aerobic bacteria the inhibitory action of valinomycin is also a function of its potassium-transporting effect which disturbs oxidative phosphorylation as a secondary consequence. In analogous fashion, valinomycin disrupts oxidative phosphorylation in the mitochondria of eukaryotic cells.

Nonactin

Another series of antibiotics known as the macrotetrolides exemplified by nonactin (Fig. 3.5) have a cyclic structure which similarly permits the enclosure of a potassium ion in a cage of eight oxygen atoms (the carbonyl and tetrahydrofuran oxygens), with the rest of the molecule forming an outer lipophilic shell. To produce this structure the ligand is folded in a form resembling the seam of a tennis ball and is held in shape by hydrogen-bonding. The action of the macrotetrolides closely resembles that of valinomycin.

Monensin

Ionophoric antibiotics of another broad group, typified by monensin (Fig. 3.7) and the closely related nigericin, carry a carboxyl group. In these

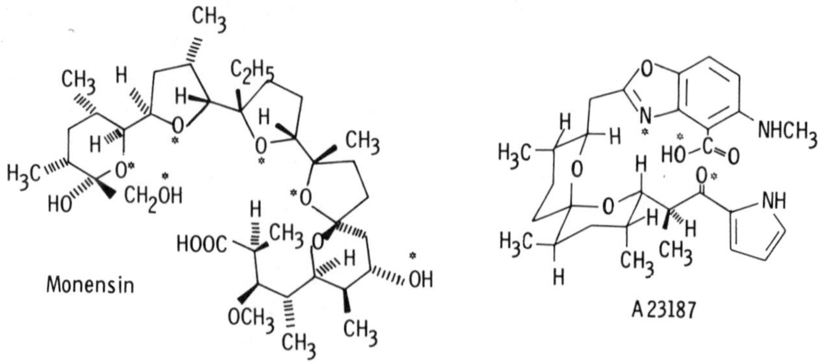

Figure 3.7 Ionophoric antibiotics. For monensin the asterisks show the six oxygen atoms which co-ordinate with the metal ion in the sodium complex. For A23187 the complex with a calcium ion involves two ligand molecules co-ordinated to the metal ion through the atoms marked with an asterisk.

compounds the molecule itself is not cyclic, but as with valinomycin a metal complex is formed in which the ion is surrounded by ether oxygen atoms and the outer surface is lipophilic. This involves a folding of the molecule which brings the carboxyl group at one end into a position where it can form strong hydrogen bonds with the alcohol groups at the other end; the structure is thus stabilized into an effectively cyclic form. Monensin binds sodium ions in preference to potassium ions, but with the closely related compound niger-icin, which allows a slightly wider spacing of the oxygen atoms, the selectiv-ity is reversed. The presence of a carboxyl group in these compounds makes an important difference to their action. They promote electrically neutral cation–proton exchange across the membrane by moving as an undissoci-ated acid in one direction and as a cation–anion complex with no net charge in the other direction. This distinguishes them from valinomycin and nonac-tin where the metal complex carries a positive charge.

These compounds have been studied mainly in artificial membranes and vesicle preparations. They promote a coupled 1 : 1 exchange of potassium or sodium ions and hydrogen ions. Since no change of charge is involved in complex formation, compounds of this type do not disturb the electrical properties of lipid-bilayer membranes.

Monensin is a compound of considerable commercial importance. It was first introduced as a coccidiostat in chickens, and has proved of exceptional utility. It has shown few signs of the development of resistance which usually terminates the effective life of drugs sold for treating coccidiosis. Later, monensin was found to improve the utilization of feedstuffs in ruminants. Its action depends on altering the balance of free fatty acid production by rumen bacteria in favour of propionate at the expense of acetate. Propionate is energetically more useful to the animal than acetate. There is also a lessening in the metabolically wasteful production of methane. The molecular basis of these effects is uncertain but the shifts in rumen metabolism can probably be attributed to differential antimicrobial actions on the complex population of micro-organisms in the rumen. The action of monensin on cell membranes is not species specific. Its lack of toxicity when given orally to farm animals probably depends upon its limited absorption from the gastrointestinal tract. The application of monensin either as a coccidiostat or as a growth promoter was strongly governed by cost and depended on the possibility of producing the compound on a large scale as cheaply as possible. This was successfully achieved by construction of fermentation vessels of unprecedented size.

The ionophores considered so far form complexes only with monovalent metal ions. A few ionophores are known which form complexes with divalent ions. Of these perhaps the most interesting is a compound known as A23187 (Fig. 3.7). This forms a 2:1 complex with calcium or magnesium ions, the calcium complex having the higher stability; it binds monovalent ions only weakly. As with monensin it is not a cyclic molecule but is able to fold into an effectively cyclic conformation by formation of a hydrogen bond between a carboxyl oxygen and the NH group of the pyrrole ring. The divalent metal ion is held in octahedral co-ordination between the polar faces of two ligand molecules. This gives an electrically neutral complex with a hydrophobic outer surface. It acts as a freely mobile carrier of these ions and causes progressive release of magnesium, uncoupling of oxidative phosphorylation and inhibition of adenosine triphosphatase in mitochondria suspended in a magnesium-free medium.

Like monensin, several divalent cationophoric antibiotics can modify rumen fermentation and improve the utilization of feedstuffs by sheep and cattle.

Gramicidin A

Gramicidin A (Fig. 3.8) (quite unrelated to gramicidin S) has many bio-chemical properties resembling those of valinomycin. It shows a specificity towards potassium ions and promotes their passage across lipid membranes. However, studies have shown that its mechanism of action is different. The most significant demonstration of this distinction depends upon measure-ments of the electrical conductivity of artificial membranes separating aqueous layers containing potassium ions. Conditions can be chosen where addition of valinomycin, nonactin or gramicidin A at 0.1 μM concentration lowers the resistance of the membrane at least 1000-fold. If the temperature is now lowered gradually the membrane reaches a transition point at which its lipid layer effectively changes phase from liquid to solid. In the presence of valinomycin or nonactin a 2°C fall in temperature at the transition point

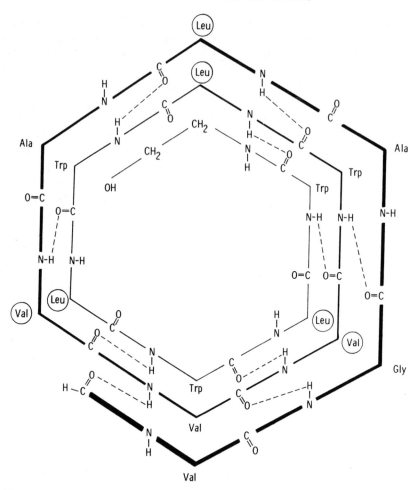

Figure 3.8 Gramicidin A. One possible helical structure having 6.3 residues per turn. Bonds drawn inwards are directed down the helix; those drawn outwards are directed up. D-Amino acid residues are circled.

causes a dramatic rise in membrane resistance, but in a similar experiment with gramicidin A resistance rises only slowly as the temperature falls. The effect with compounds of the valinomycin type is understandable since they require a liquid membrane for mobility and movement. Gramicidin A must act by some process which does not require mobility. The most obvious way is by formation of a pore which would still permit the flow of ions through a rigid membrane. Inspection of the molecular formula of gramicidin A shows it to be a linear polypeptide in which alternating amino acid residues have the L configuration. The remaining residues are either D-amino acids or glycine. The *C*-terminal is amidated with ethanolamine and the *N*-terminal carries a formyl group. The configuration allows the molecule to form an open helical structure held together by hydrogen bonds lying almost parallel to the axis of the cylinder. One possible helical form is shown in Fig. 3.8. The inside of the helix is lined with polar groups and there is a central hole about 0.4 nm in diameter. The fatty side chains of the amino acids form a lipophilic shell on the outside. One such molecule is not long enough to form a pore across a membrane, but head to head dimerization is believed to occur by bonds between the formyl groups. The existence of dimerization is supported by measurements in artificial membranes which show that conductance is proportional to the square of the concentration of gramicidin A. The length of the dimer is calculatd to be 2.5–3.0 nm which is somewhat less than the thickness of the fatty acid layer in many membranes so some distortion probably occurs during pore formation.

Conductivity measurements suggest that these pores have a transient existence, a small fraction of the antibiotic being in the form of pores at any given time. The life of a channel measured in a phosphatidylethanolamine artificial membrane was 0.35 s. However, while a pore is in existence, its transporting capacity is high. One channel is estimated to convey 3×10^7 K^+ ions s^{-1} under a potential gradient of 100 mV. Thus a low concentration of gramicidin A is a very effective carrier of potassium ions. Divalent cations are too large to traverse the gramicidin pores but block the free passage of monovalent ions.

POLYENE ANTIBIOTICS

The polyenes constitute a large group with varied molecular structures which interact with membranes in an especially interesting way. However, because the polyenes are essentially antifungal in action, they are dealt with in Chapter 6.

AN INHIBITOR OF MEMBRANE PHOSPHOLIPID BIOSYNTHESIS

Cerulenin

This compound (Fig. 3.9) is active against a range of bacteria, yeasts and fungi. Although not unduly toxic to animals, it has not been developed as an agent against microbial infections, but its unusual mode of action makes it a useful biochemical tool. It inhibits the synthesis of long-chain fatty acids by blocking 3-oxoacyl[acyl-carrier-protein] synthase. The effect is readily demonstrated by inhibition of incorporation of labelled acetate into all cell components containing fatty acids. This affects many organisms by stopping the synthesis of membrane phospholipids. In Gram-negative bacteria the outer membrane is also affected by the failure to form lipopolysaccharide and

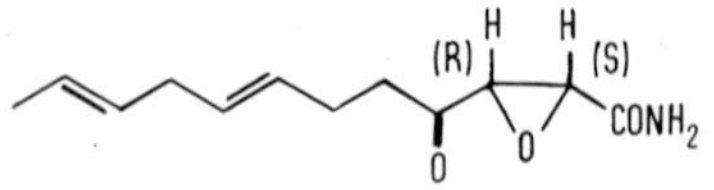

Figure 3.9 Cerulenin, an inhibitor of microbial long chain fatty acid biosynthesis.

lipoprotein. The action of cerulenin is purely bacteriostatic; inhibited cells washed free of the compound resume fatty acid production and grow normally. The inhibitory effect of cerulenin can be partly countered by providing cells with an exogenous source of long-chain fatty acids. It has also found use as a specific inhibitor of the polyketide pathway, and is used extensively in studies of the biosynthesis of this important group of compounds.

FURTHER READING

Anderson, O. S. (1984) Gramicidin channels. *Annu. Rev. Physiol.*, **46**, 531.

Bernheimer, A. W. and Rudy, B. (1986) Interactions between membranes and cytolytic peptides. *Biochim. Biophys. Acta*, **864**, 123.

Dobler, M. (1981) *Ionophores and Their Structures*, John Wiley and Sons.

Dumura, S. (1976) The antibiotic cerulenin, a novel tool for biochemistry as an inhibitor of fatty acid biosynthesis. *Bacteriol. Rev.*, **40**, 681.

Finean, J. B., Coleman, A. R. and Michel, R. H. (1984) *Membranes and Their Cellular Functions*, 3rd edn., Blackwell, Oxford.

Hugo, W. B. (1987) Mode of action of non-antibiotic antibacterial agents. In *Pharmaceutical Microbiology*, 4th edn. (eds W. B. Hugo and A. D. Russell) Blackwell, Oxford, p. 258.

Ingram, L. O. and Buttke, T. M. (1984) Effects of alcohols on micro-organisms. *Adv. Microb. Physiol.*, **25**, 254.

Jain, M. K. and Wagner, R. C. (1984) *Introduction to Biological Membranes*, John Wiley and Sons.

Lynn, B. and Hugo, W. B. (1987) Chemical disinfectants, antiseptics and preservatives. In *Pharmaceutical Microbiology*, 4th edn. (eds W. B. Hugo and A. D. Russell), Blackwell, Oxford, p. 201.

Russell, A. D. (1986) Chlorhexidine, antibacterial action and bacterial resistance. *Infection*, **14**, 212.

4 Inhibitors of nucleic acid synthesis

Many antimicrobial substances, both synthetic chemicals and natural products, directly inhibit the biosynthesis of nucleic acids. However, very few of these inhibitors are clinically useful as antimicrobial drugs because most of them do not distinguish between nucleic acid synthesis by the infecting micro-organism and that by the host. Many inhibitors of nucleic acid synthesis are therefore too toxic to the host for safe use as antimicrobial agents. However, there are a few exceptions which are described in this chapter. Some inhibitors of nucleic acid synthesis are used as anticancer agents, although their administration to patients is fraught with danger because of their lack of selectivity between normal and neoplastic cells. Not only are the dividing cancer cells attacked but also the rapidly dividing normal cells of the body. Strictly speaking, anticancer drugs fall outside the scope of this book but because several of these drugs have interesting effects on nucleic acid synthesis, a few examples are given to illustrate different modes of attack both on the biosynthesis of nucleic acid precursors and on the polymerization process.

The synthesis of DNA and the various classes of RNA is an essential function of dividing and growing cells. Inhibition of DNA synthesis rapidly results in inhibition of cell division. In addition, the biosynthesis and intercellular exchange of extrachromosomal elements of DNA in bacteria, the episomes and plasmids, are important in maintaining the flexible responses of bacteria to changes in the environment (Chapter 8).

The inhibition of RNA synthesis is followed by cessation of protein synthesis. The time elapsing between the inhibition of RNA synthesis induced by a drug such as actinomycin D and the resulting failure of protein biosynthesis may be used to indicate the rate of utilization and destruction of messenger RNA in intact cells.

Substances which interfere with nucleic acid biosynthesis can be divided into two main categories. The first group includes compounds that interfere with the elaboration of the 'building blocks' of nucleic acids, i.e. the purine and pyrimidine nucleotides. Interruption of the synthesis of any of the nucleoside triphosphates required for nucleic acid synthesis stops further macromolecular synthesis when the nucleotide precursor pool is exhausted. Many structural analogues of purines and pyrimidines and their respective nucleosides act in this way. Few such compounds have been useful as antibacterial drugs because of their lack of specificity but several purine and pyrimidine analogues used as anticancer and antiviral drugs are described later and in Chapter 6. Drugs that interfere with the biosynthesis of folic acid also inhibit nucleotide biosynthesis. Interruption of the supply of tetrahydrofolate soon brings nucleotide and nucleic acid biosynthesis to a halt.

The second group of inhibitors interferes with nucleic acid synthesis at the polymerization stage. This type of inhibition is often due to an interaction between the inhibitor and DNA and may prevent either the replication of DNA itself or interfere with the template function of DNA in RNA synthesis. Other inhibitors of polymerization hinder the function of the polymerases by direct interaction with these enzymes or the enzyme–template complexes. As we shall see, polymerase inhibitors often exhibit species specificity.

COMPOUNDS AFFECTING THE BIOSYNTHESIS OF NUCLEOTIDE PRECURSORS

Inhibitors of cofactor biosynthesis

THE SULPHONAMIDE ANTIBACTERIALS

The sulphonamides were the first compounds found to suppress bacterial infections. The original observation was made with the dyestuff Prontosil rubrum which is metabolized in the liver to sulphanilamide (Fig. 4.1). It soon became apparent that the activity was associated with the sulphonamide group. A more effective derivative of sulphanilamide was sulphapyridine, which was in turn superseded by compounds with less toxic side effects. Several of these early compounds are still in use and their structures are shown in Fig. 4.1.

Many other sulphonamide antibacterials have been developed since; these are probably no more effective than the earlier compounds, though some are much more persistent in the body and can therefore be dosed less frequently. The sulphonamides act against a wide range of bacteria, but their main success immediately following their discovery was in the treatment of streptococcal infections and pneumococcal pneumonia. Gradually the sulphonamides were displaced by naturally occurring antibiotics and their derivatives largely because of their greater potency. However they have retained a place in the treatment of urinary tract infections, some forms of meningitis and in veterinary medicine. More recently a combination therapy with trimethoprim has proved useful. The structural requirements for anti-

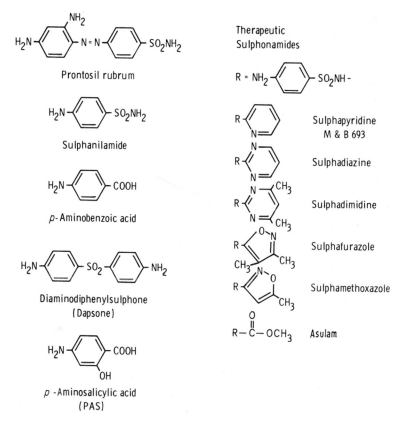

Figure 4.1 The sulphonamide antibacterials and related compounds.

bacterial activity in the sulphonamide series are relatively simple. Starting from sulphanilamide the modifications have been almost entirely variations in substitution on the nitrogen of the sulphonamide group. Substitution on the aromatic amino group causes loss of activity.

During the search for antibacterials related to the sulphonamides many compounds were synthesized and rejected because their general activity was not high enough. Among these was dapsone (Fig. 4.1) which, though it has no useful action against common infections, has an excellent effect in leprosy. It is still the drug mainly used for the treatment of this disease. Another compound which may be considered along with the sulphonamides is *p*-aminosalicylic acid (PAS) (Fig. 4.1). This substance again is not a general antibacterial agent, but has a specific action against the tubercle bacillus and is one of the trio of drugs used in the standard treatment of tuberculosis.

Both dapsone and PAS are thought to act by the same biochemical

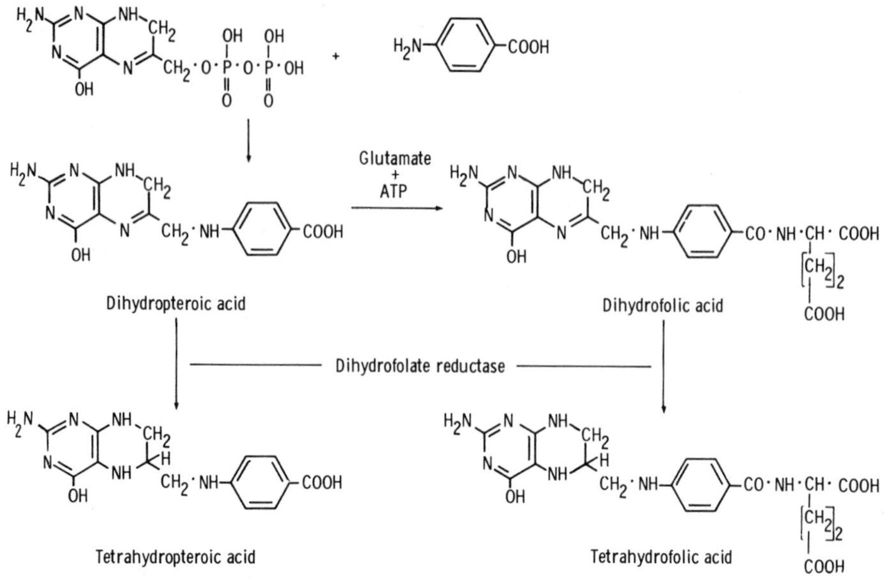

Figure 4.2 The final stages of folic acid biosynthesis.

mechanism as the sulphonamides, but the reason for their specificity in mycobacterial infections is not yet known. A few years after the discovery of the antibacterial activity of the sulphonamides, it was shown that some bacteria have a nutritional requirement for *p*-aminobenzoic acid which is involved in the biosynthesis of folic acid (Fig. 4.2). It was noted that *p*-aminobenzoate and the sulphonamides were isosteres. A competitive antagonism between *p*-aminobenzoate and these compounds on inhibition of bacterial growth was observed and this established the site of action of the sulphonamides. Later the structure of folic acid was found to contain a *p*-aminobenzyl group and its biosynthesis was shown to be inhibited by the sulphonamides. The biosynthesis proceeds to the dihydropteridine pyrophosphate derivative shown in Fig. 4.2 which then reacts with *p*-aminobenzoic acid with loss of the pyrophosphate group to give dihydropteroic acid. Sulphanilamide and other sulphonamides inhibit the reaction in an apparently competitive manner.

The sulphonamides were originally believed to compete with *p*-aminobenzoate by occupying the active site of dihydropteroate synthase without taking part in the enzymic action. More recently it has been shown that they can compete by acting as alternative substrates and bind more tightly to dihydropteroate synthase than the substrate *p*-aminobenzoate. The sulphonamides and *p*-aminobenzoate at equivalent concentrations react with the dihydropteridine pyrophosphate intermediate at similar rates. The sulpho-namide reaction products have been identified as the expected analogues of dihydropteroate. These products however probably do not play a major role in antibacterial action since they only inhibit enzymes such as dihydroptero-ate synthase at concentrations higher than those achievable in the cell.

The striking success of the sulphonamides as antibacterials, coupled with the early knowledge of their point of action, led to an extraordinary flurry of chemical activity. Every conceivable bacterial growth factor became the model for the synthesis of analogues that were hoped to repeat the success of the sulphonamides as antibacterial agents. This tremendous effort was largely fruitless. The apparently simple model provided by the antagonism of *p*-aminobenzoic acid by sulphanilamide proved completely misleading. The sulphonamides owe their effect to a set of circumstances not easily achieved elsewhere: *p*-aminobenzoate is not a metabolic intermediate in animal cells which acquire their folic acid from the diet, and the inhibition of bacterial growth by sulphonamides is not reversed by folic acid because of poor diffusion into the cells. In contrast, the sulphonamides, like *p*-aminobenzoic acid, enter the bacterial cell freely. Many biosynthetic intermediates carry phosphoric acid groups which tend to prevent their passage into bacteria from the medium, and potential inhibitors based on analogous structures share the same difficulty of access. Thus it was a number of properties fortunately coinciding that enabled the sulphonamides to succeed as anti-bacterial drugs. Since plants and protozoa also depend on the endogenous synthesis of folic acid, compounds similar to the sulphonamides have been used as herbicides, e.g. asulam (Fig. 4.1), and as antiprotozoals, e.g. sulphadoxine (Fig. 6.2).

ANTAGONISTS OF DIHYDROFOLATE REDUCTASE

When the structure of folic acid became known and its relationship to *p*-aminobenzoic acid and the sulphonamides was accepted, a search was naturally made for antagonists among structural analogues of folic acid itself. These were found, but not surprisingly they were highly toxic since folic acid derivatives, in contrast with *p*-aminobenzoic acid, play an important part in the metabolism of animal cells. The toxicity of some of these compounds towards animal cells is actually much greater than towards bacteria since the bacterial membranes are almost completely impermeable to them. The cytotoxic action of some antifolic compounds, e.g. methotrexate (Fig. 4.3), has found a practical application in the treatment of certain malignancies and psoriasis.

Although the direct analogues of folic acid were useless as antibacterial agents, other compounds with related actions have considerable importance. The potential of this type of compound was realized in two drugs developed

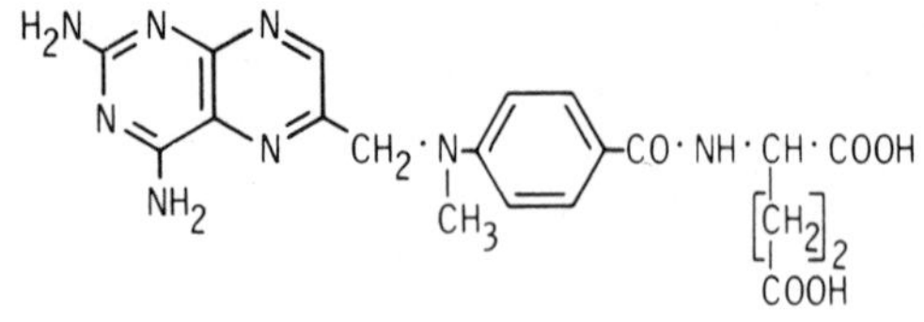

Methotrexate

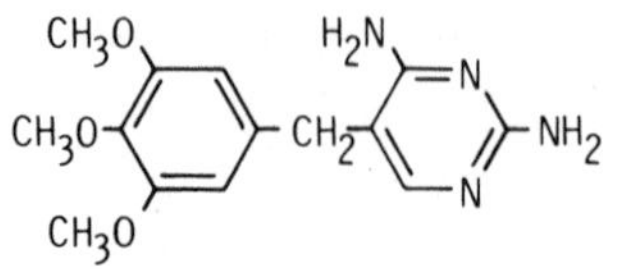

Trimethoprim

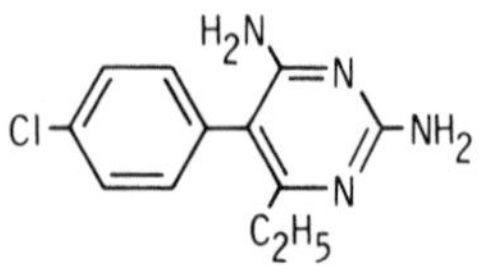

Pyrimethamine

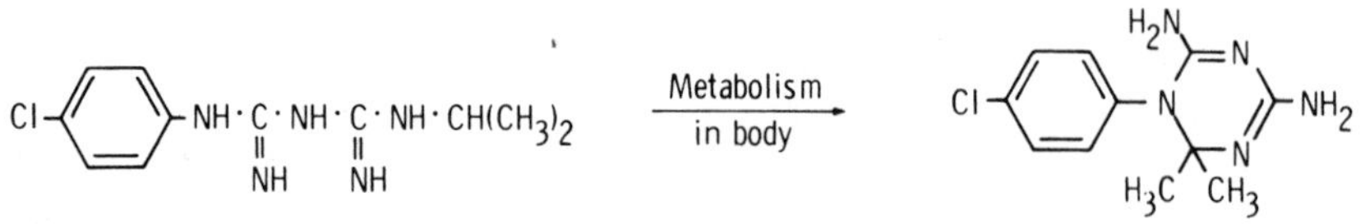

Proguanil

Figure 4.3 Chemotherapeutic agents that antagonize dihydrofolate reductase.

as antimalarials, pyrimethamine and proguanil, which is a prodrug of the active agent (Fig. 4.3).

The exact point of attack of these so-called antifolic compounds became apparent when the details of folic acid biosynthesis were fully worked out. The step leading to the production of dihydropteroic acid has already been discussed. At this point glutamic acid may be added to give dihydrofolic acid. However, in either form the dihydro compounds must be reduced to the tetrahydro state by the enzyme dihydrofolate reductase (Fig. 4.2) before they can take part as cofactors in reactions involving single carbon units. Both the cytotoxic analogues of folic acid and the antiprotozoal compounds mentioned above inhibit dihydrofolate reductase. The antimalarials were not in themselves useful antibacterial agents, but biochemical tests on these and chemically related compounds showed some interesting evidence of enzymic specificity.

Although most living cells contain dihydrofolate reductase, the enzyme evidently differs in structure from one organism to another, and a useful degree of specificity in the action of antagonists can be achieved. Thus pyrimethamine is not very active against the reductase from bacteria but it has an exceptionally strong affinity for the enzyme from *Plasmodium vinckei*; this explains its specific antimalarial action. A highly selective action against the enzyme from bacteria was eventually found in the pyrimidine derivative trimethoprim (Fig. 4.3). Reduction of the activity of bacterial dihydrofolate reductase by 50% requires a trimethoprim concentration of 0.01 μM whilst the same inhibition of the reductase from human liver requires 300 μM. The enzymes differ in several other properties; the most striking are summarized below.

	Dihydrofolate reductase	
	Mammalian	Bacterial
Substrates	Folate and dihydrofolate	Dihydrofolate only
Ks for dihydrofolate	<0.05 μM	10 μM
Effect of SH-binding agents	Inhibited	Unaffected

Trimethoprim can be used on its own as an antibacterial but is usually marketed in combination with the sulphonamide sulphamethoxazole as cotrimoxazole ('Septrin'). The combination is claimed to have a wider field of antibacterial activity than the sulphonamides and is used as an alternative to ampicillin. Both the sulphonamide and trimethoprim block the folic acid biosynthetic pathway, but at different points. This double blockage seems to be specially effective in entirely cutting off the supply of tetrahydrofolate to the bacteria.

The reduction in tetrahydrofolate levels in bacteria caused by sulphona-mides or the dihydrofolate reductase inhibitors has widespread effects on the cell. Tetrahydrofolate is required as a 'one carbon unit' donor in the biosynthesis of methionine, glycine and the formyl group of fMet-tRNA and deprivation results in reduced protein synthesis. The major effects, however, are on several stages of the biosynthesis of purine and pyrimidines which also involve C-1 compounds. The synthesis of thymine is particularly sensitive to inhibitors of dihydrofolate reductase because of the requirement for tetrahy-drofolate in the transformation of dUMP to dTMP (Fig. 4.5). When cultures of bacteria are grown in media containing amino acids and inosine, antagon-ists of folic acid synthesis cause the phenomenon known as 'thymineless death'. This effect is preventable by the addition of excess thymine or thymidine to the culture.

Inhibitors of the enzymes of nucleotide biosynthesis

Many inhibitors of the synthesis of nucleotide precursors block the elabora-tion of purine nucleotides. The complex series of reactions leading to the formation of adenine and guanine nucleoside monophosphates is summa-

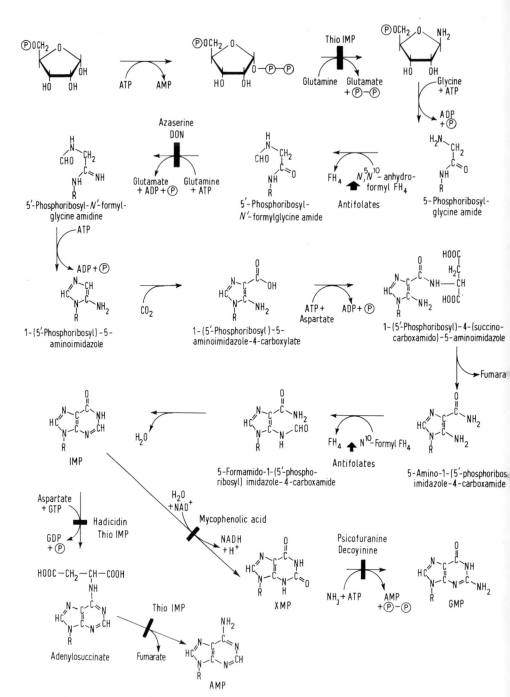

Figure 4.4 Biosynthetic pathways leading to purine nucleotides. Points of inhibition by antimicrobial compounds are shown with a bar. Interference with cofactor production is shown with an arrow. R is a phosphoribosyl group.

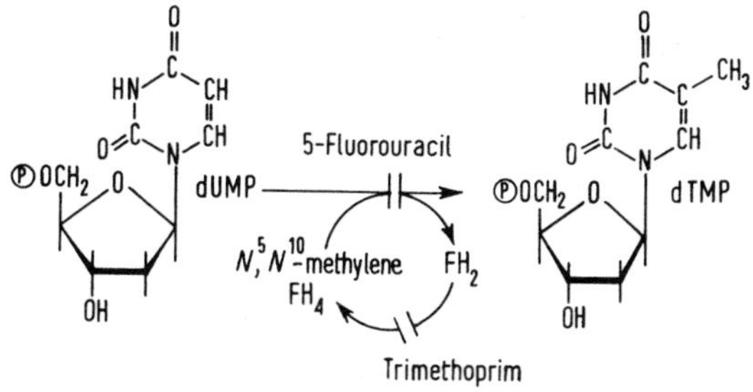

O
‖
HN—C
CH
‖
O=C   N—CH   5-Fluorouracil
℗OCH₂ O   dUMP

N^5, N^{10}-methylene   FH₂
FH₄

O
‖
HN—C   CH₃
C
‖
O=C   N—CH
℗OCH₂ O   dTMP

OH

Trimethoprim

OH

Figure 4.5 The conversion of dUMP to dTMP by thymidylate synthase and its inhibition.

rized in Fig. 4.4. We shall refer only to those steps that are relevant to the mode of action of the inhibitors.

AZASERINE AND 6-DIAZO-5-OXO-L-NORLEUCINE (DON)

Both these antibiotics are produced by streptomycetes and are structurally similar. They can be regarded as analogues of glutamine (Fig. 4.6). Azaserine and DON inhibit the growth of many micro-organisms, but because they are also toxic to mammalian cells they have only been used clinically as antitumour drugs.

Both drugs prevent the attachment of glutamine to its binding site on the enzyme which converts formylglycine amide ribonucleotide to the corresponding amidine. Azaserine combines irreversibly with the SH group at the active centre. DON probably reacts similarly and is known to combine with

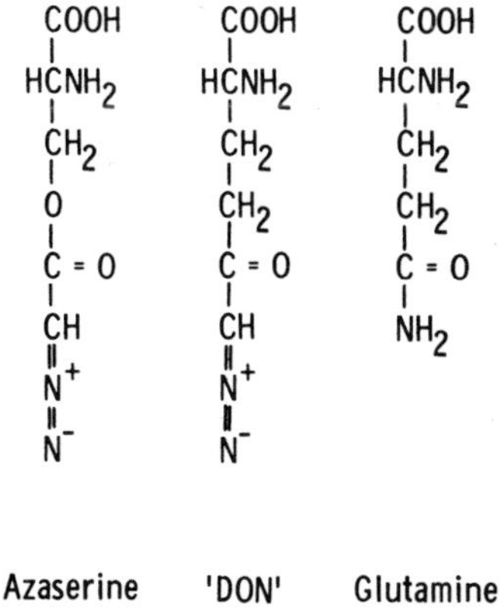

COOH
HCNH₂
CH₂
O
C = O
CH
N⁺
N⁻

COOH
HCNH₂
CH₂
CH₂
C = O
CH
N⁺
N⁻

COOH
HCNH₂
CH₂
CH₂
C = O
NH₂

Azaserine   'DON'   Glutamine

Figure 4.6 The structural similarities between the antibiotics azaserine and diazo-oxonorleucine (DON) and the amino acid glutamine.

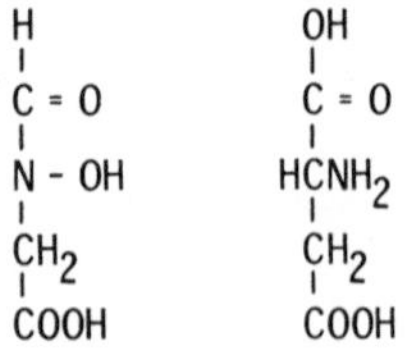

```
        H                  OH
        |                  |
        C = O              C = O
        |                  |
        N - OH             HCNH₂
        |                  |
        CH₂                CH₂
        |                  |
        COOH               COOH

      Hadacidin      L - Aspartic acid
```

Figure 4.7 The structures of hadacidin and L-aspartic acid.

the SH group at the glutamine-binding site of CTP synthetase, the enzyme that catalyses the transfer of an $-NH_2$ group from glutamine to UTP. Despite the similarities between azaserine and DON, there are also important differences. For example, DON is more potent than azaserine as an inhibitor of purine nucleotide biosynthesis, while azaserine has radiomimetic and mutagenic properties that are lacking in DON.

HADACIDIN

A product of many species of *Penicillium*, hadacidin (*N*-formylhydroxyaminoacetic acid) is an analogue and antagonist of L-aspartic acid (Fig. 4.7). Hadacidin has both antimicrobial and anticancer activity, although its clinical effectiveness against human malignant disease is disappointing. Hadacidin is a competitive inhibitor of adenosylsuccinate synthase and inhibits the conversion of IMP to adenosylsuccinate. The K_i for hadicidin is 4.2 μM whereas the K_m for L-aspartate is 150 μM. Hadicidin only significantly competes with L-aspartate in this reaction; it has little or no antagonistic activity against L-aspartate in other reactions, e.g. the conversion of 1-(5'-phosphoribosyl)-5-aminoimidazole-4-carboxylate to the corresponding amide requires aspartate but is unaffected by hadicidin. Again hadicidin only weakly competes with aspartate in pyrimidine biosynthesis.

PSICOFURANINE AND DECOYININE

Psicofuranine is a naturally occurring antibiotic with a nucleoside type of structure (Fig. 4.8). It has both antimicrobial and anticancer activity. However, its use as an anticancer drug has been limited by a tendency to cause pericarditis in man.

Despite its structural resemblance to adenosine, psicofuranine inhibits nucleic acid synthesis in bacteria by blocking the final stage in the biosynthesis of GMP, i.e. the amination of XMP. The enzyme that catalyses this reaction in *Escherichia coli* is inhibited non-competitively by psicofuranine and

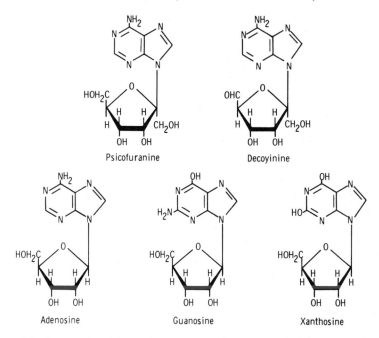

Figure 4.8 Two nucleoside antibiotics,, psicofuranine and decoyinine compared with usual purine nucleosides.

also by the closely related antibiotic decoyinine (Fig. 4.8). This inhibition is reversed by adenosine, which itself may inhibit XMP aminase non-competitively. The inhibition by psicofuranine, however, is not reversed by either xanthosine or guanosine. Certain reagents, such as urea and 2-mercaptoethanol, 'desensitize' XMP aminase to the action of psicofuranine and decoyinine and yet preserve the catalytic function of the enzyme. This suggests that these antibiotics do not act at the active site of the enzyme but at an allosteric site that provides an inhibitory control mechanism in the native enzyme. Desensitization may modify the allosteric site so that it no longer binds the inhibitory nucleoside derivatives. The inhibition of XMP aminase by adenosine presumably fulfils some control function in purine nucleotide metabolism.

MYCOPHENOLIC ACID

This is an antibiotic of considerable antiquity which was first detected in 1896 as a product of *Penicillium stoloniferum*. Chemical studies eventually revealed its chemical structure some 50 years later (Fig. 4.9). Although the antibacterial activity of mycophenolic acid is very limited, its activity against

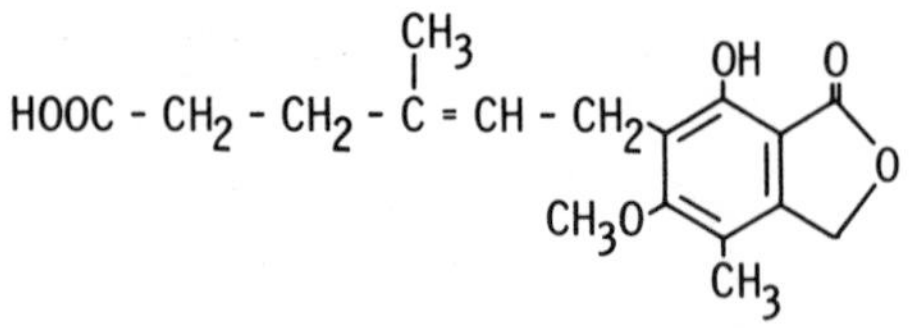

$$\text{HOOC - CH}_2\text{ - CH}_2\text{ - } \overset{\overset{\displaystyle CH_3}{|}}{C}\text{ = CH - CH}_2$$

Figure 4.9 Mycophenolic acid.

a range of fungi is more striking. It also has excellent activity against many experimental tumours in rodents. Unfortunately the compound proved to be inactive against human cancers, apparently because it is rapidly metabolized in man to an inactive product.

The antimitotic action of the antibiotic depends mainly on its inhibition of the enzyme which converts IMP to XMP, i.e. IMP dehydrogenase (IMP–NAD^+ oxidoreductase). This depletes the supply of guanine nucleotides essential for DNA and RNA synthesis. Mycophenolic acid does not obviously resemble IMP or NAD or indeed any other reactant in purine nucleotide metabolism. Possibly it induces conformational distortions in the enzyme that hinder its normal functioning. Prolonged storage of a preparation of IMP dehydrogenase from *Bacillus subtilis* leads to a partial loss of sensitivity of the enzyme to mycophenolic acid, suggesting that the inhibitor does not act directly at the active site.

6-MERCAPTOPURINE

This antitumour drug inhibits several interconversions between purine nucleotides. The compound enters the metabolic pathways and becomes converted to the nucleotide analogue 6-thioinosinic acid (thioIMP) which inhibits several enzymes (see Fig. 4.4) either as a direct inhibitor, e.g. IMP dehydrogenase, or as an allosteric feedback inhibitor, e.g. in the production of 5-phosphoribosylamine. Mercaptopurine is also converted to thioguanine-containing nucleotides which are incorporated into DNA and affect transcription to RNA.

5-FLUOROURACIL AND 5-FLUORODEOXYURIDINE

These compounds, which are clinically important antitumour drugs, are metabolized to 5-fluorodeoxyuridine monophosphate which is a potent inhibitor of thymidylate synthase. Inhibition of this enzyme restricts the availability of thymidine for incorporation into DNA. The antifungal drug 5-fluorocytosine is similarly metabolized to a compound capable of inhibiting DNA synthesis in yeasts (see Chapter 6).

INHIBITORS OF NUCLEIC ACID SYNTHESIS AT THE POLYMERIZATION LEVEL

In the final stage of nucleic acid synthesis the nucleoside triphosphates are added to a growing polynucleotide chain in which the constituent nucleosides are joined by 3'-5'-phosphodiester linkages. The nucleoside triphosphates react sequentially under the influence of the appropriate polymerase. The biosynthesis of the polynucleotide proceeds in a 5' to 3' direction. The sequence in which the nucleotides are added is, of course, governed strictly by the base-pairing requirements of the template strand. In DNA synthesis, the initiation of the new strand also requires a primer, i.e. a strand terminating in a free 3'-OH group on which the first nucleotide can substitute. RNA synthesis on the other hand does not require a primer for strand initiation. During the replication of double-helical DNA both strands are copied in an antiparallel manner by a mechanism of continuous replication of the 'leading' strand and discontinuous in the other. In RNA synthesis, however, only the so-called 'plus' strand of the DNA acts as a template.

Inhibitors of polynucleotide chain formation act either by disturbing the template function of the nucleic acid or by direct inhibition of the specific polymerase. Although DNA is a chemically unreactive molecule, the need to preserve its conformation and continuity for great lengths makes it vulnerable to agents which bind to it or introduce covalent modifications. These interactions cause profound changes in the physical properties of the macromolecule, the consequences of which are incorrect transcription to RNA or impairment of DNA replication – both effects potentially lethal to the cell. Inhibition of polymerase function may be an indirect result of the interaction between the inhibitor and the template molecule; inhibition of polymerase activity can also result from a direct attack of the inhibitor on the enzyme or on the specific subcellular location of the template–enzyme complex. In bacteria, at least, the subcellular site for DNA replication is thought to be on the cytoplasmic membrane.

Compounds which interact with template DNA

Compounds which interact with DNA fall into two categories, those that interact non-covalently with DNA and those that form covalent bonds with DNA. A summary of compound types illustrating both types of interaction is shown in Table 4.1.

INTERCALATING COMPOUNDS

Many rigidly planar polycyclic molecules interact non-covalently with double-helical DNA by a mechanism known as intercalation. The planar

Table 4.1 Inhibitors of nucleic acid synthesis that interact with DNA

Compounds	Interaction	Resultant effects
(a) NON-COVALENT INTERACTIONS		
Phenanthridines, e.g. ethidium	General intercalation	Inhibit fidelity of DNA replication and RNA elongation
Acridines, e.g. proflavine	General intercalation	″
Actinomycin D	Intercalation at GC pairs, projects into minor groove	Inhibit RNA synthesis > DNA synthesis
Quinoxalines, e.g. echinomycin	Double intercalation at CpG regions, projection into minor groove	Inhibit RNA elongation > DNA
Anthracyclines, e.g. daunorubicin	Intercalation at AT base-pairs	Inhibit DNA initiation and elongation, RNA initiation
Netropsin	Non-intercalative at AT regions, 3–5 base-pair interaction	Prevents binding of DNA polymerase
(b) COVALENT INTERACTIONS		
Bleomycins	Chain breakage	Inhibit RNA and DNA synthesis
Mitomycin	Strand cross-linking	Interferes with DNA synthesis
Anthramycin	Reacts with guanine residue in minor groove	Inhibits DNA synthesis
Neocarzinostatin	Interacts in AT-rich areas, four base-pair region, also strand breakage	Inhibits initiation of DNA synthesis

character of such molecules enables them to insert or intercalate between the adjacent stacked base-pairs of the double helix. To permit this insertion, there must be a preliminary local unwinding of the double helix to produce spaces between the stacked pairs into which the planar polycyclic molecule can move. The much reproduced diagram in Fig. 4.10 shows schematically how polycyclic structures may intercalate between the stacked base-pairs. The hydrogen-bonding between the base-pairs remains undisturbed, although there is some distortion of the smooth coil of the sugar phosphate

Figure 4.10 Diagram to represent the secondary structure of normal DNA (left) and DNA containing intercalated molecules (right); the stacked bases are separated at intervals by the intercalators (black), resulting in some distortion of the sugar–phosphate backbone of the DNA. [This diagram is reproduced by kind permission of Dr L. Lerman and the Wistar Press, *J. Cell. Comp. Physiol.*, **64** Suppl. 1 (1964)].

backbone as the intercalated molecules maintain the double helix in a partially unwound configuration. It is believed that this partial unwinding prevents normal replication or transcription of the DNA.

The details of a specific drug–DNA interaction depend largely on the structure of the drug, and several examples are described below. The intercalation of a molecule into DNA is detectable by various physical changes. DNA solutions show an increase in viscosity; there is a decrease in sedimentation coefficient and an increase in thermal stability. The extent of these changes is proportional to the amount of drug intercalated into the double helix.

The increase in the viscosity of DNA solutions treated with intercalating drugs is explained by the restricted degree of irregular tertiary coiling that the double helix can undergo. The DNA–drug complex is, therefore, both straighter and stiffer than the uncomplexed nucleic acid and these changes raise the viscosity. The reductions in sedimentation coefficient and buoyant density of DNA following intercalation result from a reduction in the mass per unit length of the nucleic acid. For example, a proflavine molecule (Fig. 4.11) increases the length of the DNA by about the same amount as an extra base-pair, but because proflavine has less than half the mass of the base-pair, the mass per unit length of the complexed DNA is decreased. The increased thermal stability of intercalated DNA is probably due in part to the extra energy needed to remove the bound molecule from the double helix in addition to that required to separate the strands. Undoubtedly, electrostatic interactions between the intercalated molecule and the two DNA strands assist stabilization of the double helix.

In certain tumour viruses and bacteriophages, in the kinetoplasts of try-panosomes and in bacteria harbouring sex factors and drug-resistance plasmids (see Chapter 8) double-stranded DNA exists as covalently closed

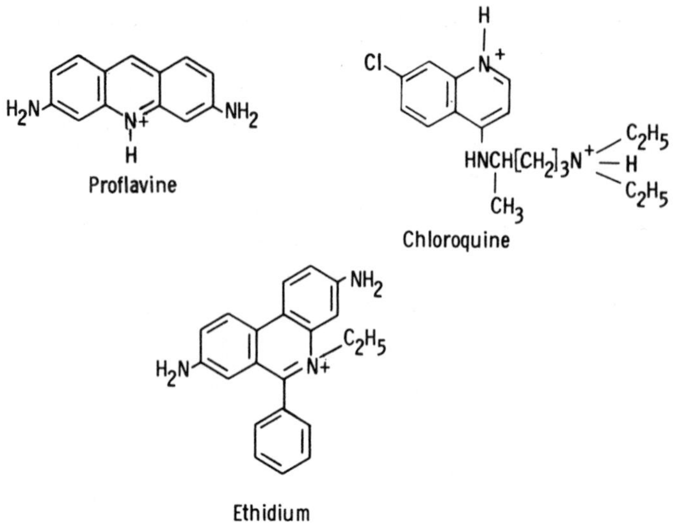

Figure 4.11 Three molecules that intercalate with DNA.

circles. Isolated circular DNA that is covalently closed via the usual 3′–5′-phosphodiester bond is characteristically supercoiled. These supercoils occur because the circular molecule is in a state of strain. The strain is relieved and the supercoils often disappear when single-stranded breaks or 'nicks' are produced by the action of an endonuclease.

Closed circular DNA has an unusual affinity for intercalating molecules which, because they partially unwind the double helix, also reduce the supercoiling of the DNA. If the unwinding proceeds beyond a certain point, as more and more drug is added, the DNA starts to adopt the supercoiled form again, except that the supercoiling occurs in the opposite direction from that of the uncomplexed DNA. At this point the affinity of the closed circular DNA for the intercalated molecules actually declines until it is less than that of nicked DNA.

The diminished affinity of closed circular DNA for ethidium (Fig. 4.11) at high concentrations of the drug permits a convenient separation of closed circular DNA from nicked DNA, as the sedimentation coefficient and buoyant density of DNA with a lower content of intercalated material are significantly higher. This effect has been invaluable in the isolation of closed circular DNA on a preparative scale.

It is also possible that the initial higher affinity of supercoiled DNA for intercalating molecules may in part account for their peculiar specificity of action against organelles and organisms believed to contain circular DNA.

Treatment of bacteria with acridines may lead to the disappearance of plasmids from the cells.

The mitochondria of certain strains of yeast are severely and irreversibly damaged by growth in the presence of ethidium apparently owing to a drug-induced cytoplasmic mutation affecting the mitochondrial DNA. The kinetoplast of trypanosomes is also seriously affected by intercalating agents, DNA synthesis in this organelle being selectively inhibited. Eventually the kinetoplast disappears altogether. As this adversely affects the life cycle of trypanosomes it is possible that the selective attack on the kinetoplast may underlie the trypanocidal activity of certain intercalating drugs.

Acridines, phenanthridines and chloroquine

The medical history of the acridine dyes extends over some 70 years since proflavine (Fig. 4.11) was used as a-topical disinfectant on wounds during the First World War. Proflavine is too toxic to be used as a systemic antibacterial agent, but the related acridine, mepacrine (Fig. 1.2), found wide application as an antimalarial drug. Chloroquine (Fig. 4.11) is still an important antimalarial agent and the phenanthridine compound ethidium (Fig. 4.11) is a useful trypanocide.

The compounds all bind to the nucleic acids of living cells and the phenomenon forms the basis of vital staining, since the nucleic acid–dye complexes exhibit characteristic colours when examined by fluorescence microscopy. The dyes also bind readily to nucleic acids *in vitro* and the visible absorption spectra of the ligand molecules undergo a metachromatic shift to longer wavelengths.

Two types of binding to DNA are recognized: a strong primary binding which occurs in a random manner in the molecule, and a weak secondary binding. The strong primary binding occurs only with DNA, although many other polymers bind the dyes by the secondary process. Primary binding to DNA, which is mainly responsible for the ability of these drugs to interfere with nucleic acid synthesis, causes changes characteristic of intercalation. The location of the intercalated molecule may be stabilized by electronic interactions, between its fused rings and the heterocyclic bases of the DNA above and below it. The complex may also be stabilized by hydrogen-bonding between the amino groups of the proflavine and ethidium and the charged oxygen atoms of the phosphate groups in the sugar–phosphate backbone. In the case of chloroquine the projecting cationic side chain may form a salt linkage with the phosphate residue. The intercalation of molecules into DNA disturbs the fidelity of replication and prevents the correct RNA polymerase–DNA interaction and consequently inhibits the initiation rather than the extension of the RNA strand.

Actinomycin D

This complex antibiotic (Fig. 4.12) was discovered in 1940. It is an exceedingly toxic substance and consequently has no therapeutic value as an antimicrobial agent. The main interest in actinomycin D lies in its unique interaction with DNA and its application as a specific inhibitor of DNA-dependent RNA synthesis in cell biology. Actinomycin D also has a limited clinical use as an anticancer agent particularly in the treatment of a renal tumour of children known as Wilms' tumour.

The interaction between DNA and actinomycin D is readily demonstrated: addition of DNA causes a prompt change in the absorption spectrum of actinomycin D solutions. The addition of RNA, however, has no such effect. Furthermore, addition of DNA to solutions of biologically inactive derivatives of actinomycin D does not change their spectra. The following structural features of the actinomycin D molecule are involved in the interaction with DNA. (i) The intact cyclic pentapeptide lactones are indispensable. Both in solutions of actinomycin and in crystalline complexes of the drug with deoxyguanosine and oligonucleotides, the peptidolactones extend laterally from the tricyclic chromophore, rather like the arms of the letter T (Fig. 4.13). The amino acid sequence is apparently not critical, although replacement of the L-*N*-methylvaline with L-valine abolishes the ability to interact with DNA. (ii) Monomethylation of the amino group of the chromophore causes a 90% loss of activity; other substitutions or removal of the amino group abolish activity entirely. (iii) The quinonoid oxygen is also essential for activity.

The DNA must also have certain essential features for it to interact with actinomycin D. (i) The DNA must contain guanine–cytosine pairs, although the related base 2,6-diaminopurine can substitute for guanine in synthetic

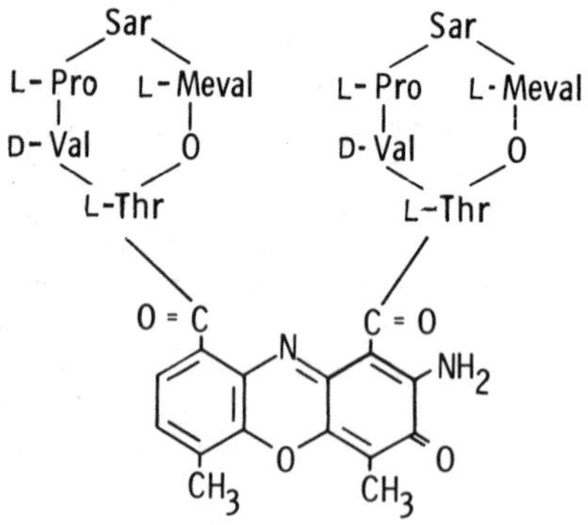

Figure 4.12 Actinomycin D. Sar, sarcosine; Meval, *N*-methylvaline.

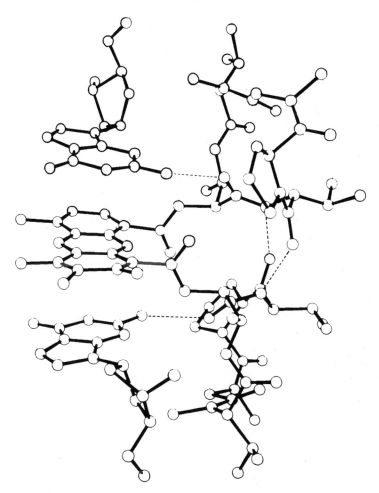

Figure 4.13 Computer-drawn illustration derived from the X-ray crystallographic analysis of the complex of one molecule of actinomycin D with two molecules of deoxyguanosine. The complex is viewed from one side and the chromophore of actinomycin D can be seen sandwiched between the two purine rings. The dotted lines represent the hydrogen bonds which link the pentapeptide rings of the antibiotic to each other and those which link the guanine 2-amino groups with the carbonyl oxygens of the L-threonine residues of the antibiotic. Additional weaker hydrogen bonds, which are not shown, are also proposed between the guanine N_3-ring nitrogens and the NH groups of the L-threonines. The diagram is reproduced with the kind permission of Dr H. M. Sobell and the editors of *Nature New Biology* [*Nature (London) New Biol.*, **231**, 200 (1971)].

DNA polymers. (ii) The DNA must be a double-stranded helical structure; the small degree of binding that occurs between actinomycin D and heat-denatured DNA is probably due to limited helical regions formed as the DNA is rapidly cooled. (iii) The sugar moiety must be deoxyribose; double-stranded RNA (containing guanine) does not interact with actinomycin D to any significant extent.

Detailed X-ray studies of a crystalline complex formed between one molecule of actinomycin D and two molecules of deoxyguanosine have revealed that the guanine residues are stacked above and below the chromophoric ring of actinomycin D (Fig. 4.13). The drug molecule is stabilized internally by hydrogen-bonding between its two cyclic pentapeptides; two hydrogen bonds are formed between the NH of the D-valine in one ring and the CO of the D-valine in the other ring. The two cyclic pentapeptides are thus held in such a way that strong hydrogen-bonding can occur between the 2-amino groups of the guanine residues and the CO groups of the L-threonines. Weaker hydrogen bonds connect the ring nitrogen in the 3-position of the guanines with the NH groups on the same L-threonines. Hydrophobic bonding is also thought to occur between the deoxyribose residues and the isopropyl groups of the N-methylvalines.

Extending this model to the complex between actinomycin D and DNA, the cyclic pentapeptides are thought to lie in the minor groove of the double helix. Additional hydrogen-bonding could occur between the 2-amino group of the drug chromophore and a phosphate oxygen and also the sugar–ring oxygen of a deoxycytidine residue. In a duplex DNA molecule the guanine residues stacked above and below actinomycin D are base-paired with cytosine residues.

Bifunctional intercalating drugs: the quinoxaline antibiotics

The quinoxaline antibiotics are complex molecules exemplified by echinomycin and triostin A (Fig. 4.14) which are produced by several streptomycetes. They are powerfully cytotoxic molecules, active against Gram-positive bacteria but not against Gram-negative organisms probably because they fail to penetrate the outer envelope. Since they are very toxic to dividing mammalian cells, the quinoxalines have had some limited use as experimental anticancer agents in man. Their cytotoxic action has been traced to an interaction with DNA. The binding of each echinomycin molecule extends the double helix by 0.63 nm, which is almost twice the distance that accommodates a single intercalated aromatic chromophore.

This extension occurs because echinomycin intercalates both of its quinoxaline rings into DNA. The other physical changes in DNA molecules brought about by interaction with echinomycin, i.e. reduction in sedimentation coefficient, removal and reversal of supercoiling, are also consistent with

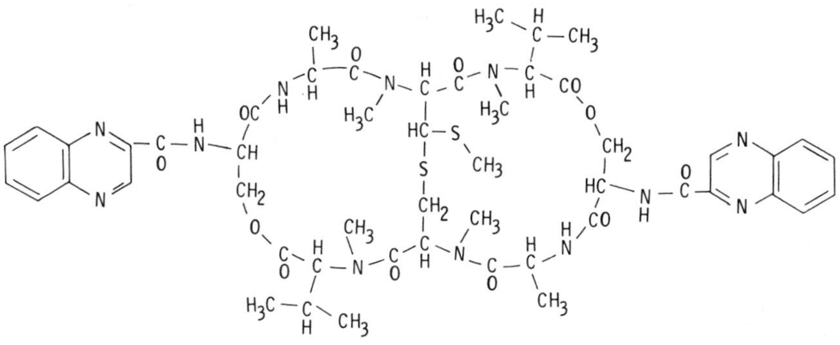

Echinomycin

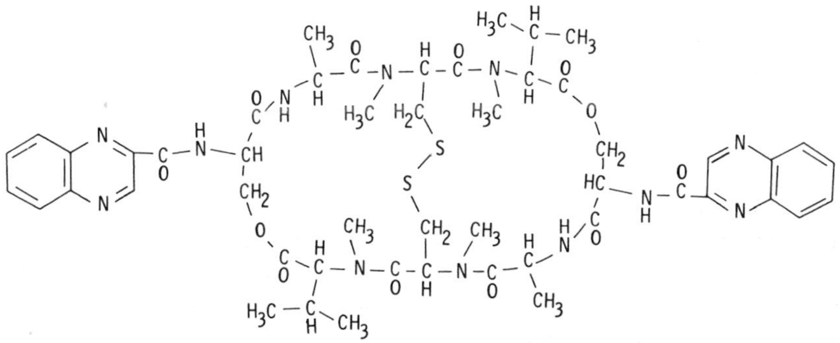

Triostin A

Figure 4.14 The quinoxalines; echinomycin and triostin A.

intercalation. The peptide ring of echinomycin is rigid and quite planar and the whole molecule can be considered as a rigid disc with the planar quinoxaline rings at opposite ends extending in the same direction perpendicular to the plane of the disc. This unique structure enables the insertion of the twin quinoxaline residues into the double helix at locations approximately 1 nm, i.e. two DNA base-pairs, apart. The integrity of the octapeptide ring must be essential for intercalation because the dimethyl ester of echinomycinic acid, in which the ring is ruptured, cannot interact with DNA. An intact cross-bridge in the peptide ring is also essential, though it can vary in length and constitution, as exemplified in triostin A (Fig. 4.14) which is as potent as echinomycin as a bifunctional intercalator.

Binding to DNA tends to be tighter than with simple (monofunctional) intercalators and there is evidence for nucleotide sequence selectivity. 'Footprinting' experiments have been employed to identify preferred ligand-binding sites in natural DNA fragments (CpG sequences in the case of echinomycin) and reveal that local perturbations of the helical structure can be propagated into DNA regions flanking the antibiotic-binding sites. Crystallographic evidence suggests that echinomycin recognizes GC base-pairs by forming hydrogen bonds between the carbonyl groups of alanine residues in the antibiotic and the 2-amino groups of guanine nucleotides in the minor groove of the DNA helix. Kinetic studies support the hypothesis that sequence-selective antibiotic molecules 'shuffle' between different binding sites in the process of locating their optimal (preferred) sites.

Anthracycline-containing antibiotics

The anthracycline glycosides include the daunorubicin types, mithramycin and nogalamycin (Fig. 4.15). These natural products have planar fused rings which are able to intercalate between base-pairs. The interaction is influenced by hydrogen-bonding of the molecules to the polymer backbone. X-ray diffraction analysis and enzyme inhibition studies indicate that nogalamycin prevents the *de novo* synthesis of the alternating co-polymer poly(AT), by *Escherichia coli* DNA polymerase. Model building shows that the molecule protrudes into both the minor and major grooves in DNA. Binding to poly(AT) is much faster than to poly(GC). Adriamycin (Fig. 4.15), another member of this family, shows significant activity against a variety of neoplasms including leukaemia, lung cancer and sarcomas. In addition to intercalation adriamycin binds covalently to DNA, probably in a similar way to mitomycin (see later).

The intercalating compounds inhibit both DNA synthesis and DNA-dependent RNA synthesis in intact cells and in cell-free preparations. Some compounds, such as actinomycin D and ethidium, inhibit RNA synthesis at concentrations much lower than those needed to inhibit DNA synthesis. Others such as proflavine interfere equally with DNA and RNA synthesis. Since strand separation is an essential for the replication of DNA, the increased stability of the double helix following intercalation may be a major factor in the inhibition of DNA replication. In the inhibition of DNA-dependent RNA synthesis, the intercalated molecules probably hinder the attachment of RNA polymerase to the DNA template so that the number of initiations of new RNA molecules is markedly decreased. The elongation of the nucleic acid chain is also inhibited by the presence of intercalated molecules in the template.

One possible explanation for the selective action of actinomycin D against RNA synthesis is that the RNA polymerase finds its normal progression along the minor groove blocked by the presence of the cyclic pentapeptide

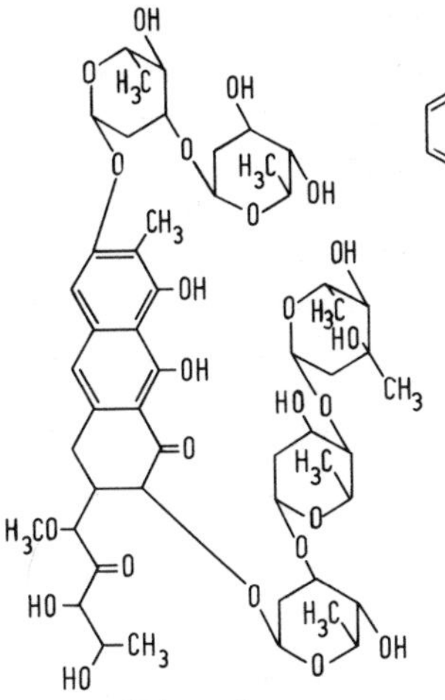

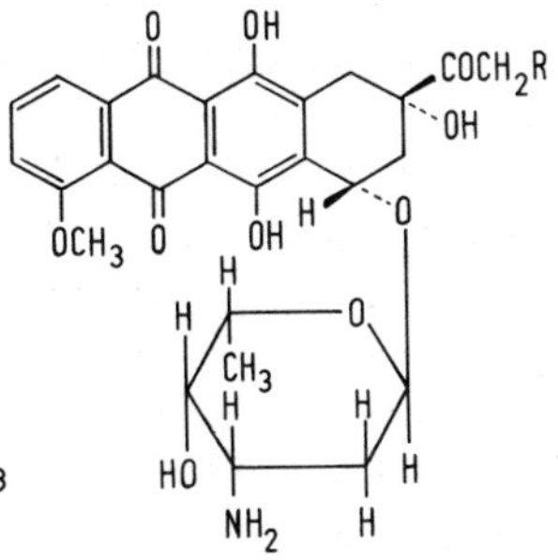

Daunorubicin : R = H
Doxorubicin (Adriamycin) : R = OH

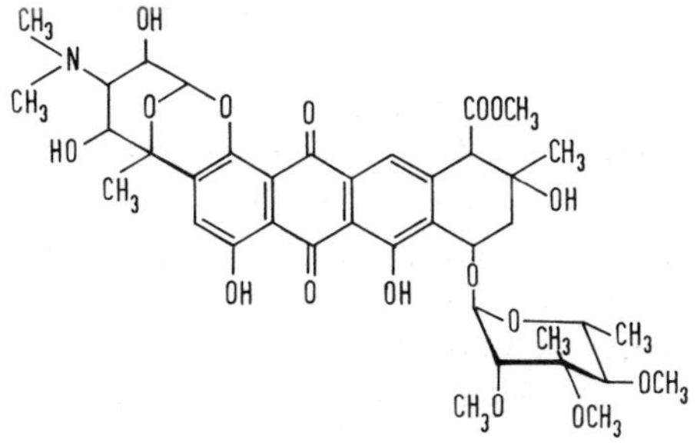

Mithramycin

Nogalamycin

Figure 4.15 Three sugar-containing compounds that intercalate DNA.

rings. This may also be true for other agents such as the quinoxalines. The inhibition of DNA synthesis caused by higher concentrations of actinomycin D may be associated with the greater stability of the double helix brought about by an increased number of drug molecules bound to the DNA.

NON-INTERCALATING COMPOUNDS

Netropsin

Netropsin (Fig. 4.16) is a basic oligopeptide antibiotic which forms extremely stable complexes with duplex DNA especially with base pairs in AT-rich

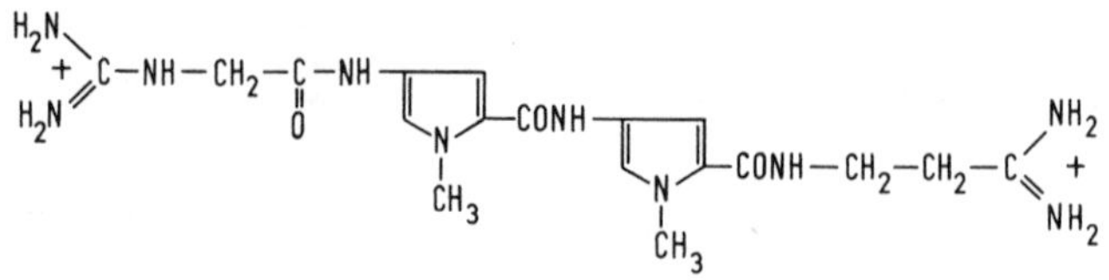

Figure 4.16 Netropsin.

regions of DNA. The binding is non-intercalative but depends on hydrogen-bonding and electrostatic interactions. The interaction is thought to extend over three to five base-pairs and inhibits the effective binding of polymerases to the DNA.

DRUGS WHICH INTERACT COVALENTLY WITH DNA

Bleomycins

The bleomycins comprise a family of complex, closely related antibiotics (Fig. 4.17) which react covalently with DNA. These compounds are toxic to mammalian cells as well as to Gram-positive bacteria and so their medicinal use is confined to anticancer treatment. We describe them here because their biochemical action on DNA is unique.

The bleomycins produce single- and double-stranded breaks in DNA molecules and release the DNA bases. Ferrous iron, which forms a tight complex with bleomycin, molecular oxygen and magnesium ions are required for the strand-breaking reaction. This reaction resembles a catalytic process, since the bleomycin molecule is used repeatedly to produce chain scissions. The chemical mechanism of the action of bleomycins on DNA is uncertain but the first step may involve partial intercalation of the bis-thiazole ring system, possibly adjacent to thymine. Since the strand-breaking action of bleomycin would normally obscure the characteristic changes caused by intercalation, experiments were specially designed to minimize strand breakage. It could then be shown that bleomycin relaxes the super-coils of closed circular DNA and that linear DNA is lengthened by 0.31 nm by each intercalating molecule of bleomycin. The drug forms a covalent bond with thymine at the 2-carbonyl position. The attached bleomycin can then degrade DNA by a reaction which is thought to involve the conversion of the complexed Fe(II) to Fe(III) and the release of either hydroxyl radicals or superoxide anions, which bring about the elimination of the neighbouring thymine residue and eventually scission of the DNA chains. The strand-breaking bleomycins are so destructive to the DNA template that both DNA and RNA synthesis rapidly fail after drug treatment.

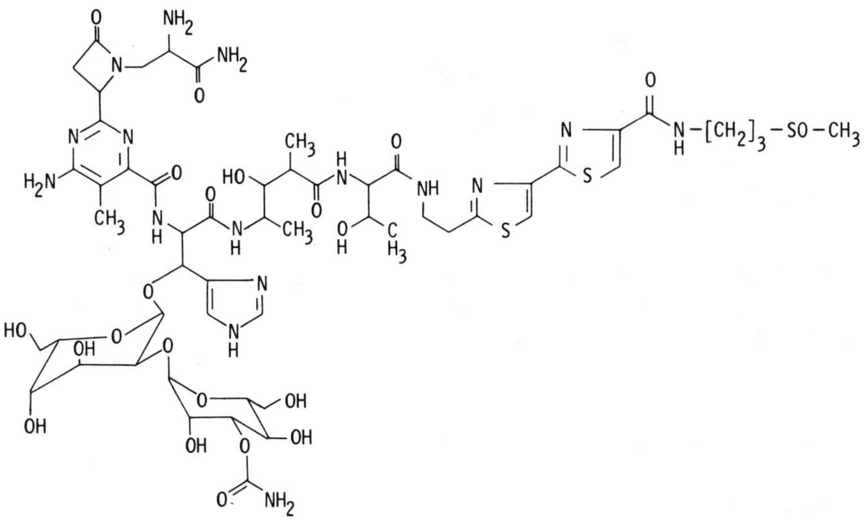

Figure 4.17 Bleomycin A$_1$, a member of a family of antibiotics which 'catalyse' chain breakage in DNA.

Mitomycin and porfiromycin: cross-linkers of DNA

Mitomycin C and porfiromycin (Fig. 4.18) belong to a series of chemically related substances that are produced by various species of streptomycetes. They are powerful antimitotic agents and rapidly cause the death of both microbial and mammalian cells. Their medical use is confined to anticancer therapy. The biocidal activity of these compounds depends upon a highly specific interaction with DNA which cross-links the DNA strands.

Continuous monitoring of the optical absorbance at 260 nm of solutions of normal double-stranded DNA during heating reveals an abrupt increase of absorbance as the hydrogen bonds linking the complementary base-pairs of the two strands are ruptured. Once separated, the complementary polynucleotide chains lose their correct alignment in relation to each other. Recovery of the native double-helical structure only occurs when the cooling process is carried out very slowly (this is true for bacterial and viral DNA; renaturation is rarely observed with mammalian DNA). If the cooling process is rapid, complete renaturation is prevented, since correct alignment of the complementary base-pairs cannot occur. However DNA isolated from bacteria treated with mitomycin-like compounds shows a very different pattern of behaviour. The absorbance profile is essentially normal as the temperature is

Figure 4.18 Mitomycin C (R = H) and porfiromycin (R = CH₃). Enzymically generated hydroquinones are thought to be converted to the quinone methides which are then subject to nucleophilic attack by purine bases of the DNA.

raised. Even during rapid cooling, however, extensive renaturation of the DNA is observed.

This property of DNA from mitomycin-treated cells is due to the existence of heat-stable covalent cross-links between the complementary strands. Although the hydrogen bonds are ruptured as the DNA is heated, the cross-links remain intact and hold the two strands in register. On cooling, therefore, hydrogen-bonding between correct base-pairs is facilitated and the double helix readily re-forms (Fig. 4.19).

Addition of mitomycin to solutions of DNA *in vitro* does not lead to the formation of cross-links. Metabolic activation of the drug is required. There is an initial NADPH-dependent enzymic reduction of mitomycin to the corresponding hydroquinone followed by reactions that lead to the formation of the reactive quinone methide shown in Fig. 4.18. The quinone methide is then subjected to nucleophilic attack at two positions by guanine bases in both strands of DNA thus cross-linking the double helix.

A preliminary step to the covalent interaction with DNA may be inter-

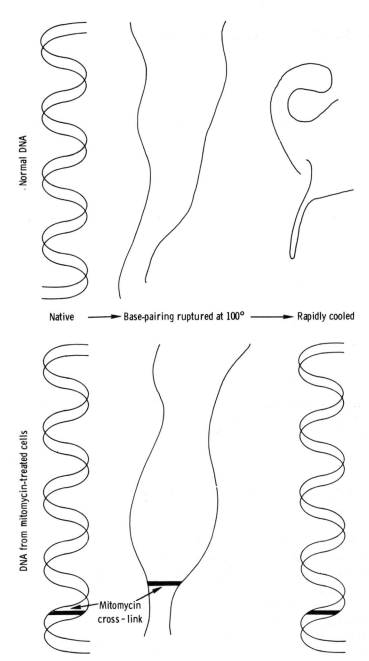

Native ⟶ Base-pairing ruptured at 100° ⟶ Rapidly cooled

Normal DNA

DNA from mitomycin-treated cells

Mitomycin cross-link

Figure 4.19 Heat denaturation of solutions of DNA isolated from normal bacteria is not reversible on rapid cooling, as correct alignment of complementary base-pairs of the separated strands cannot occur under these conditions. The covalent cross-link in DNA from mitomycin-treated bacteria, however, holds the strands together during heating and greatly facilitates correct base-pairing and hence extensive renaturation on rapid cooling.

calation of the ring system into the double helix. This accounts for the highly specific nature of the complex formed with DNA. Mitomycin-induced cross-links in DNA molecules are relatively rare, not exceeding 1 per 1000 nucleotide pairs. Some mitomycin molecules must also alkylate one or other of the DNA chains without forming a cross-link, as there is good evidence that there is about ten times more mitomycin bound to DNA than can be accounted for by the number of cross-links.

Most of the biochemical effects of the mitomycins on cells can be explained by their characteristic interaction with DNA. DNA synthesis stops whenever the replicating fork reaches a mitomycin cross-link. The degradation of DNA which eventually follows the treatment of cells with mitomycins is due to the excision of the cross-linked zones and to the appearance of nucleases associated with lysogenic phages induced by mitomycins. The synthesis of RNA and protein ceases some considerable time after the inhibition of DNA synthesis by the mitomycins. The transcription of DNA progressively fails as the alkylated DNA is degraded.

Anthramycin

The antitumour activity of anthramycin (Fig. 4.20) is attributable to its ability to interact with DNA and the consequent inhibition of nucleic acid synthesis. Anthramycin forms a labile covalent attachment to DNA. The stability of this bond depends on the secondary structure of the DNA and can be broken *in vitro* by denaturation of the DNA. Anthramycin reacts with a guanine residue in double-stranded DNA and this non-intercalative interaction results in thermal stabilization of the duplex and occurs in the minor groove. The interaction limits the 'breathing' movements allowed for the chains. Analysis by ^{13}C NMR spectra demonstrated that the covalent interaction occurred at the C-11 position of anthramycin.

Neocarzinostatin

This antitumour natural product belongs to a family of protein antibiotics that selectively inhibit DNA synthesis. Initiation is inhibited. The holoanti-

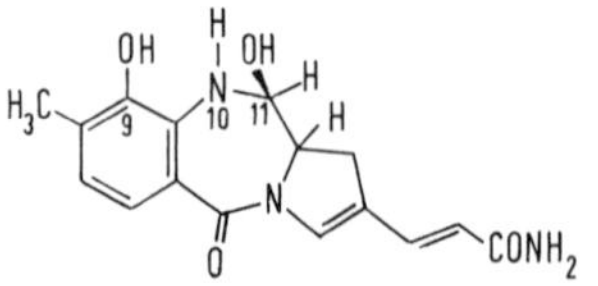

Figure 4.20 Anthramycin, an antitumour antibiotic that reacts covalently with guanine residues in DNA.

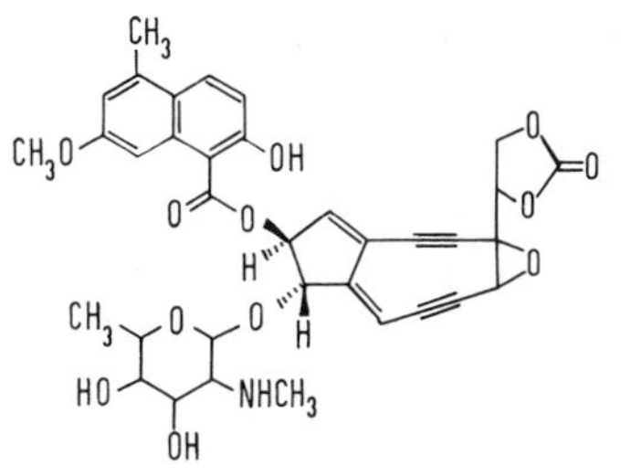

Figure 4.21 The chromophore of neocarzinostatin.

biotic consists of a non-protein chromophore (Fig. 4.21) tightly bound to an apoprotein of molecular weight 10 700. The protein protects the labile chromophore from degradation. The chromophore binds specifically to DNA, especially to regions rich in A and T, with a tight binding region consisting of four base-pairs. The chromophore contains a naphthoic acid part, an amino sugar and an unsaturated link unit. The naphthoic subunit intercalates DNA and the positively charged amino sugar binds electrostatically to the negatively charged sugar phosphate backbone of DNA. These two regions anchor the middle piece to the deoxyribose of a thymidylate residue. Upon activation of the drug with a thiol, in the presence of oxygen, there is a selective oxidation of the 5'-C of deoxyribose to produce a DNA strand break with a phosphate at the 3'-end and a nucleoside 5'-aldehyde at the other. Strand breaks occur almost entirely at thymidylate and adenylate residues.

Inhibitors of nucleic acid polymerases

All the compounds so far considered which inhibit DNA polymerization interact with the template in some way. Few of these compounds are useful antimicrobials. The reason for this is the lack of selectivity between bacterial and mammalian cells. This is what would be expected since mammalian and bacterial DNAs are essentially similar. However, there are a number of useful antimicrobial agents which specifically inhibit microbial nucleic acid polymerases, whilst having essentially no effect on the corresponding mammalian enzymes. In addition some compounds, which are antitumour agents, have little effect on bacterial systems.

RNA POLYMERASE INHIBITORS

The rifamycins (Fig. 4.22) comprise a group of closely related antibiotics produced by *Streptomyces mediterranei*. They have good activity against Gram-positive bacteria and *Mycobacterium tuberculosis* but are much less active against Gram-negative organisms, apparently because of the low permeability of these organisms to rifamycins. Chemically the rifamycins are closely related to streptovaricin (Fig. 4.22) and the two groups of antibiotics appear to have a similar mode of action. They strongly inhibit RNA synthesis in sensitive bacteria and also in cell-free extracts by binding to and inhibiting DNA-dependent RNA polymerase. The drugs neither bind to nor inhibit the corresponding mammalian enzyme. Bacterial mutants have been isolated which are highly resistant to rifampicin (a member of the rifamycin group) and the RNA polymerase from these cells is resistant to the antibiotic and shows little affinity for it.

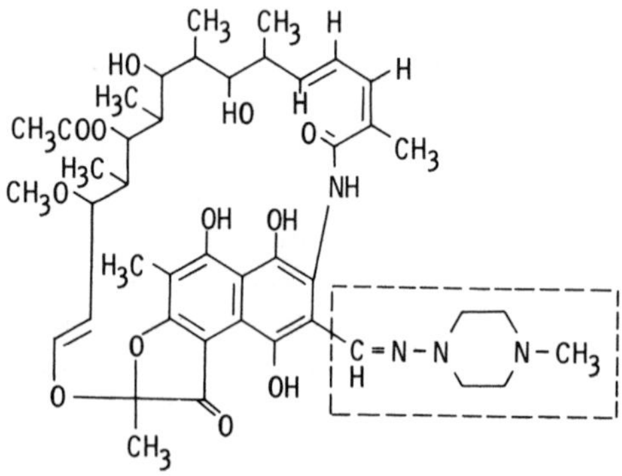

Rifampicin

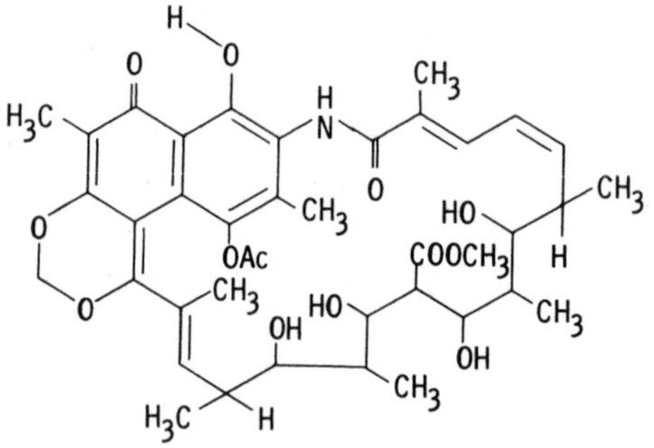

Streptovaricin D

Figure 4.22 Two antibiotics that selectively inhibit bacterial DNA-dependent RNA polymerase. Rifampicin is a semi-synthetic member of the rifamycin group; the synthetic side chain is enclosed by the dotted line. Streptovaricin D is related in structure to the rifamycins; jointly the rifamycins and streptovaricins are known as ansamycins.

Bacterial DNA-dependent RNA polymerase consists of two major components, one called the minimal or core enzyme and the other the sigma factor. The sigma factor is essential to the normal initiation of RNA synthesis *in vivo*, although it can be dispensed with *in vitro* when certain templates such as mammalian DNA are used. The efficiency of RNA synthesis in the absence of sigma factor is usually much reduced. When DNA from the T4 bacteriophage is used as a template, there is an absolute requirement for sigma factor for initiation *in vitro*. The sigma factor may be involved in the recognition of specific initiation sites on a selected strand of the DNA, enabling the RNA polymerase to attach itself to the template. Rifampicin interferes with the initiation of RNA synthesis, although the antibiotic does not prevent the binding of the intact polymerase to the DNA template.

Despite its key role in initiation, the sigma factor is not the target for rifampicin action, since studies with rifampicin-sensitive and -resistant polymerases indicate that rifampicin resistance is associated with the core enzyme. The core enzyme has been resolved into α, β, β_1 and ω subunits, and rifampicin forms a tight one-to-one complex with the β subunit. The ability to form such a complex is missing in the β subunit isolated from a rifampicin-resistant polymerase. Although the drug is strongly bound to the β subunit, the bonding is not covalent and there may be hydrophobic interactions between the fused aromatic ring system of rifampicin and aromatic amino acids of the enzyme. The binding of rifampicin to the enzyme is a two-stage process:

$$R + E \rightleftharpoons RE \rightleftharpoons RE^*$$

The first stage is a fast bimolecular reaction followed by a second slower unimolecular process involving a conformational change in the enzyme that is necessary for the inhibitory action of rifampicin. The overall dissociation constant for the interaction is very low, 3 nM. Rifampicin has little effect on the formation of the first phosphate diester bond in the polymerization process, but its inhibition of the synthesis of the next bond is virtually total and the drug interferes with the binding of an incoming purine nucleoside triphosphate to the active site of the enzyme. However, once the initiation of RNA chain synthesis progresses beyond the second or third phosphate diester bond further chain elongation is insensitive to the action of rifampicin.

Mitochondrial RNA synthesis in yeast and virtually all eukaryotes is insensitive to rifampicin but not to a hydrophobic octyloxime derivative, rifampicin AF-013. This and other hydrophobic derivatives of rifampicin can inhibit RNA-directed DNA synthesis by tumour viral polymerases. At high concentrations these derivatives inhibit not only certain mammalian RNA polymerases, but also several other related enzymes, and their value as specific reagents is therefore questionable.

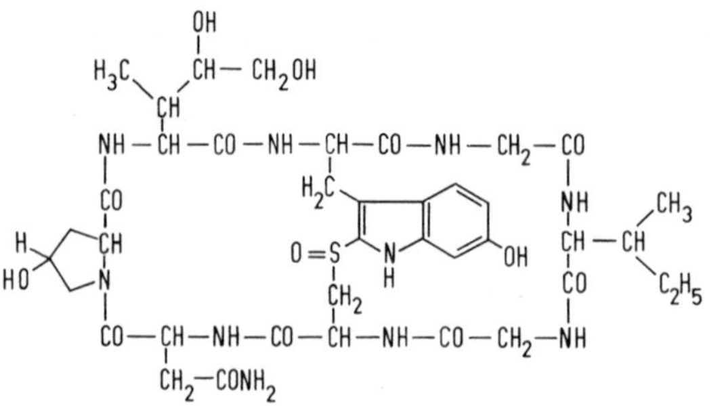

Figure 4.23 Two inhibitors of RNA polymerases.

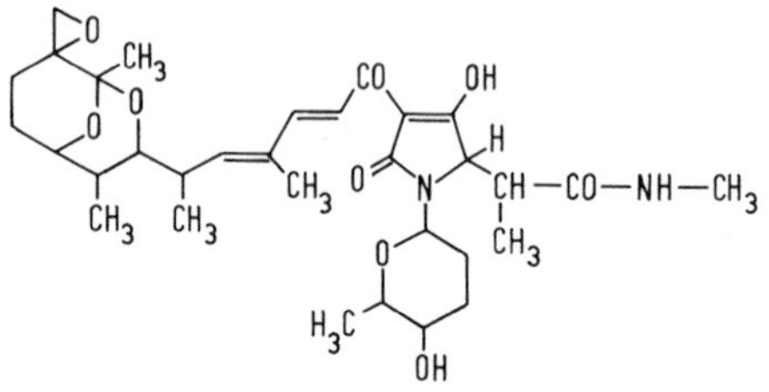

Streptolydigin (Fig. 4.23), like the rifamycins and streptovaricins, is a specific inhibitor of bacterial RNA polymerase. However, it inhibits chain elongation as well as the initiation process. Streptolydigin increases the stability of the purified RNA polymerase–DNA template complex while rifampicin prevents the stabilization induced by streptolydigin.

The β subunit of the polymerase core enzyme bears the streptolydigin-binding site and the increased stability of the enzyme–template–antibiotic complex delays the progress of the enzyme along the template without affecting the accuracy of the transcriptional process.

Despite the evidence obtained *in vitro* for the mode of action of streptolydigin, studies of its effects on intact *Escherichia coli* cells indicate that streptolydigin *in vivo* may accelerate the termination of RNA chains. The rate of elongation of RNA chains is unaffected but streptolydigin may destabilize the transcription complex *in vivo*, thus permitting premature attachment of termination factors. Only more research will resolve this apparent conflict between studies *in vitro* and *in vivo* on the mode of action of streptolydigin.

The sulphur-containing bicyclic polypeptide α-amanitin (Fig. 4.23) is a potent compound from the fungus *Amanita phalloides* which specifically inhibits mammalian RNA synthesis. The DNA-dependent RNA polymerase (polymerase II) responsible for mRNA synthesis is the most sensitive target.

The polymerase in mammalian cell nuclei which synthesizes 5S RNA and tRNA (polymerase III) is fairly insensitive, whilst polymerase I which makes other RNAs is almost totally insensitive. Bacterial RNA polymerase is unaffected by α-amanitin. The selective inhibition of the mammalian type II enzyme indicates specific binding to the enzyme and ultracentrifugation and equilibrium dialysis experiments reveal a 1:1 complex with the enzyme's 140 000 molecular weight subunit. Inhibition of RNA biosynthesis occurs after initiation of transcription has begun, indicating that α-amanitin specifically inhibits chain elongation.

DNA POLYMERASE INHIBITORS

The hydroxyphenylazopyrimidines, e.g. 6-(p-hydroxyphenylazo)uracil (Fig. 4.24), are potent antibacterial agents, but they have not been used clinically, mainly because their activity is limited to Gram-positive organisms.

The antibacterial action of these compounds depends on their ability to inhibit DNA synthesis in sensitive cells. DNA synthesis in Gram-negative organisms is unaffected. The inhibitory action on DNA synthesis has been analysed in detail and has the following features. (i) Hydroxyphenylazopyrimidines must first be reduced to the corresponding hydrazines (Fig. 4.24) before they exert their inhibitory effect. This reduction occurs metabolically in bacterial cells. (ii) Of three DNA polymerases that can be isolated from *Bacillus subtilis*, only polymerase III is inhibited by reduced hydroxyphenylazopyrimidines. This enzyme effects the semi-conservative replication of DNA, whilst polymerase I is concerned solely with repair. A mutant of *Bacillus subtilis*, resistant to hydroxyphenylazopyrimidines contains DNA polymerase III which is also resistant to these drugs. *Escherichia coli*, which is intrinsically resistant to the hydroxyphenylazopyrimidines, contains three DNA polymerases that correspond to those from *Bacillus subtilis*. None of the *Escherichia coli* enzymes is inhibited by the pyrimidine derivatives. (iii) The inhibition of *Bacillus subtilis* DNA polymerase III by reduced 6-(p-hydroxyphenylazo)uracil (HPUra) is competitively and specifically antagonized by dGTP, while the inhibition by reduced 6-(p-hydroxyphenylazo)isocytosine (HPIso) is relieved by dATP but not by dGTP. These results, together with a study of the nuclear magnetic resonance spectra of the reduced hydroxyphenylazopyrimidines in the presence and absence of various purine and pyrimidine bases, nucleosides and nucleotides, indicate that the inhibitors interact with both the polymerase and the DNA template. The interaction with DNA depends on an unusual base-pairing arrangement in which HPUra pairs with cytosine and HPIso with thymine. This base-pairing prevents the access of purine deoxynucleoside triphosphate substrates to their appropriate sites on the template. Complex formation between the drugs and DNA has an absolute require-

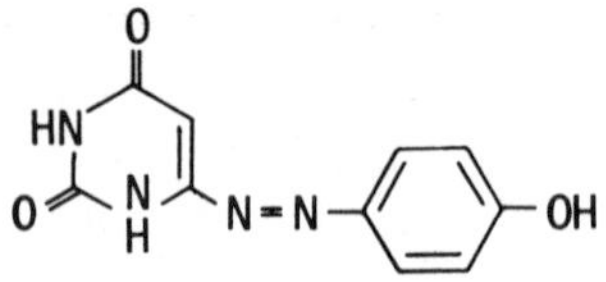

6-(*p*-Hydroxyphenylazo) uracil

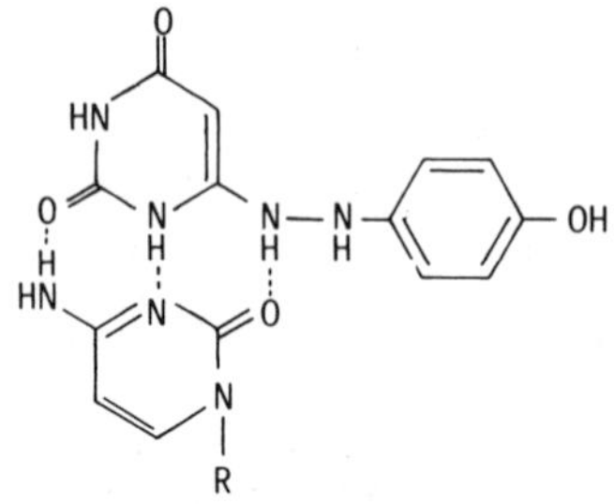

Figure 4.24 6-(*p*-Hydroxyphenylazo)uracil, an example of a group of simple synthetic chemicals which inhibit polymerase III from Gram-positive bacteria. The bottom diagram shows a possible mechanism of base-pairing of the metabolically reduced form of the compound with a cytosine residue of the DNA template. R is the remainder of the nucleotide residue.

ment for double-stranded DNA with a base-paired 3′-OH primer strand terminus and a distal unpaired pyrimidine in the template strand with which the drugs can base-pair. The 3′-OH terminus of the primer strand apparently gives the complex considerable stability. The aromatic moiety of a hydroxyphenylazopyrimidine is thought to participate in a stacking interaction with an aromatic amino acid of DNA polymerase III in a hydrophobic site close to the active site (see Fig. 4.25).

Analogues substituted in the para position of the aryl ring with a hydrophobic group inhibit eukaryotic DNA polymerases with low K_i values, e.g. 50 nM for yeast polymerase I. These compounds are, however, less potent against bacterial enzymes. Although the unique sensitivity of DNA polymerase III of Gram-positive bacteria remains an intriguing problem, the hydroxyphenylazopyrimidines have nevertheless proved valuable in highlighting previously unsuspected differences in DNA synthesis among bacterial genera.

In eukaryotic cell systems three enzymes which synthesize DNA have been identified. DNA polymerase α is the principal enzyme involved in DNA synthesis, as shown by use of the tetracyclic diterpenetetraol aphidicolin (Fig. 4.26). This compound inhibits the growth of mammalian cells, but not those of yeasts which are impermeable to it. The growth inhibition is brought

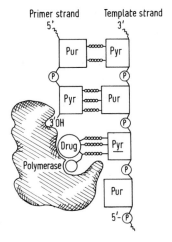

Figure 4.25 A proposed mechanism of action for the hydroxyphenylazopyrimidines, showing the drug binding to the DNA template and preventing the polymerase action. This diagram is reprinted by kind permission of Drs N. C. Brown, G. E. Wright and MacMillan Press (*Drug Action at the Molecular Level*, ed. G. C. K. Roberts, (MacMillan, 1977).

about by a specific action on polymerase α; aphidicolin has no effects on the β or γ polymerases or on bacterial DNA polymerases.

Most kinetic analyses of the mechanism of inhibition of DNA polymerase α by aphidicolin have been carried out with DNA as template. These indicate that inhibition is competitive with respect to dCTP, non-competitive with respect to the other nucleoside triphosphates and uncompetitive with respect to DNA. However, when DNA synthesis by isolated nuclei or intact cells was examined, inhibition was competitive for all the nucleosides together, but non-competitive for individual nucleosides. When aphidicolin is allowed to bind to DNA polymerase enzyme before the interaction with the nucleotide substrate the complex becomes catalytically inactive. Aphidicolin

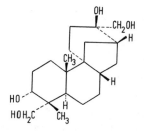

Figure 4.26 Aphidicolin: an inhibitor of eukaryotic DNA polymerase.

probably does not bind directly to the nucleotide sites but prevents substrate binding as a result of interaction with overlapping sites or inducing interactions between sites. Why the discrepancy exists between purified enzyme and native enzyme of intact cells or nuclei is unclear but it suggests that the microenvironment in the intact nucleus is different from that for the isolated enzyme.

Animal cell lines resistant to aphidicolin either produce increased levels of polymerase α or have aphidicolin-insensitive enzyme. The presence of aphidicolin-binding sites on polymerase α of various eukaryotic species suggests that the structure of these regions may have been extensively conserved.

TOPOISOMERASE INHIBITORS

The view that DNA exists in bacterial cells as an extended double-helical molecule becomes untenable when the topology is considered. The length of bacterial DNA is around $1300\ \mu$m and typically the cell into which it fits is around $1\ \mu$m in diameter. Clearly there must be a high degree of ordered quaternary structure in the DNA to allow its accommodation in the cells. This is achieved by supercoiling the DNA. Special enzymes and proteins induce torsional stresses into the molecule. The enzymes thereby alter the three-dimensional shape of DNA whilst maintaining the primary structure and the genetic information encoded on it. These enzymes are also involved in DNA replication and transcription. When a circular supercoiled DNA molecule is replicated the two daughter molecules would become interlocked without a means of removing the supercoils. The number of links would equal the original number of supercoils, around 3000 in a bacterium, and separation of the progeny would be impossible. There are, however, enzymes which remove the supercoils to allow the separation of daughter chromosomes, and then subsequently catalyse supercoiling to facilitate efficient

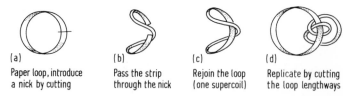

(a) Paper loop, introduce a nick by cutting

(b) Pass the strip through the nick

(c) Rejoin the loop (one supercoil)

(d) Replicate by cutting the loop lengthways

Figure 4.27 A schematic representation for the catenation of circular DNA molecules. The reader is encouraged to try this experiment. Take a strip of paper (a), introduce one complete twist into it (one supercoil) by passing the strip through a break (b) and rejoining (c). Cut down the length of the strip as if in semi-conservative replication. The result is two 'daughter molecules' which are interlinked (d). If two supercoils are introduced then a more complex linkage can be seen.

packaging of DNA. The so-called catenation of chromosomes can perhaps be more easily understood by performing the simple model experiment shown in Fig. 4.27. Similarly when DNA is transcribed into RNA it is essential for the DNA to undergo localized swivelling movements – otherwise the RNA polymerase and the growing RNA chain would have to revolve around the DNA template.

The enzymes which facilitate these topological changes are known as topoisomerases. They induce a single- or double-strand break in the DNA and permit the removal or insertion of a supercoil either by allowing one strand to rotate round the other before it is rejoined (topoisomerase I) or by permitting a double strand to pass through a double nick before rejoining (topoisomerase II). The nicked ends are covalently bound to the enzymes via tyrosine residues. Both enzymes alter the superhelicity of the DNA. For more detailed information a specialized text should be consulted.

DNA gyrase is a prokaryotic topoisomerase II enzyme, which is unique in that it introduces negative supercoils into DNA. All other known topoisomerases remove supercoils. DNA gyrase possesses two types of sub-unit, one of which facilitates the 'make and break' reactions (A) the other being an ATPase which provides the energy to allow the enzyme to catalyse these changes (B). Inhibition of this enzyme prevents the topological changes needed by the cell for DNA replication and RNA transcription. Two types of compound inhibit gyrase activity, each of which act on different subunits. (1) The 4-quinolone antibacterials which are exemplified by the relatively simple compound nalidixic acid (Fig. 4.28). This agent is mainly active against Gram-negative bacteria and is used in the treatment of urinary tract infections. The introduction into the molecule of the 6-fluoro and 5-piperidine groups increases potency greatly and such compounds, e.g. ciprofloxacin (Fig. 4.28), have some activity against Gram-positive organisms. The quinolones interact with the A subunit which is involved in the strand breakage and inhibit the activity of DNA gyrase by freezing the enzyme–DNA complex in the 'break' position. This allows an accumulation of double-strand nicks in the chromosome and causes cell death. (2) The B subunits of the enzyme are inhibited by the naturally occurring hydroxy-

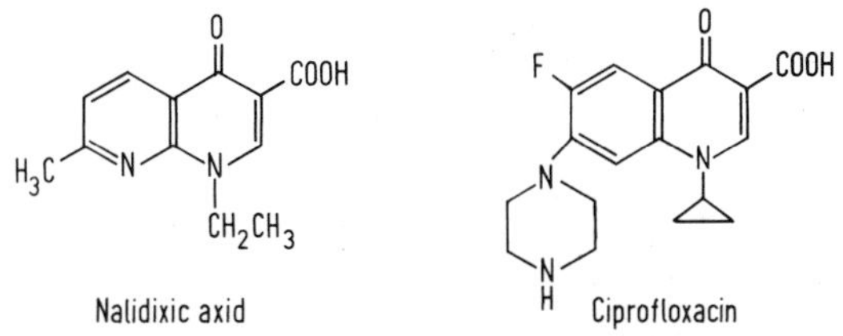

Figure 4.28 Quinolone antibacterial inhibitors of DNA gyrase.

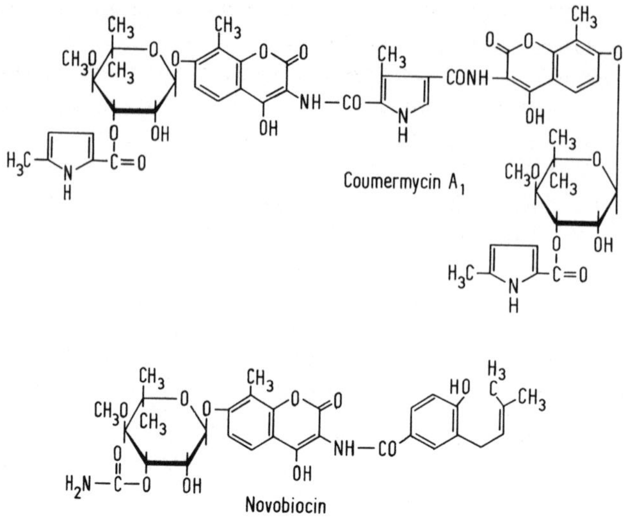

Figure 4.29 Hydroxycoumarin DNA gyrase inhibitors.

coumarins novobiocin and coumermycin (Fig. 4.29). These compounds block supercoiling activity by competing for the ATP-binding site on the subunit. ATP is normally hydrolysed by the B subunit to allow the enzyme to return to its activated state after it has introduced a single supercoil into a DNA double helix. The hydroxycoumarins are much more potent against Gram-positive organisms than Gram-negatives due to differences in uptake of the drugs. Coumermycin is about fifty times more potent against *Escherichia coli* than novobiocin although its molecular weight of about 1400 is twice that of novobiocin. The high potency and large molecular size of coumermycin suggest it may have a specific entry mechanism into bacterial cells.

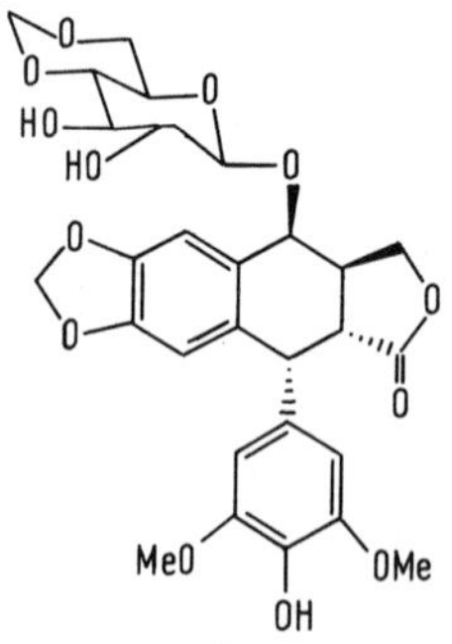

Figure 4.30 Antitumour agent etoposide.

Despite its lower potency novobiocin has some clinical use as a reserve drug when other agents fail to cure an infection.

In mammalian systems there are topoisomerase II enzymes which do not introduce torsional stress, but allow a reduction in supercoiling, presumably to allow transcription and replication of the DNA. The anticancer agent etoposide (Fig. 4.30) inhibits this enzyme. The inhibition, like the inhibition of gyrase by 4-quinolones, results in double-strand breakage in the DNA which leads to cell death.

FURTHER READING

Brown, N. C., Dudycz, L. W. and Wright, G. E. (1986) Rational design of substrate analogues targeted to selectively inhibit replication of specific DNA polymerases. *Drugs Exp. Clin. Res.*, **12**, 555.

Fox, K. R. and Waring, M. J. (1986) Footprinting reveals that nogalamycin and actinomycin shuffle between DNA binding sites. *Nucleic Acid Res.*, **14**, 2001.

Hitchings, G. H. (1983) Inhibition of folate metabolism in chemotherapy. *Handb. Exp. Pharmacol.*, **64**.

Kornberg, A. (1980) *DNA Replication*, W. H. Freeman.

Loeb, L. A., Liu, P. K. and Fry, M. (1986) DNA polymerase: enzymology, function, fidelity and mutagenesis. *Progr. Nucleic Acid Res. Mol. Biol.*, **33**, 58.

Moore, H. W., Czerniak, R. and Hamdam, A. (1986) Natural quinones as quinomethide precursors – ideas in rational drug design. *Drugs Exp. Clin. Res.*, **12**, 475.

Ogawa, T. and Okazaki, T. (1980) Discontinuous DNA replication. *Annu. Rev. Biochem.*, **49**, 421.

Record, M. T., Mazur, S. J., Melancon, P., Roe, J. H., Shaner, S. L. and Unger, L. (1981) Double helical DNA conformations, physical properties and interactions with ligands. *Annu. Rev. Biochem.*, **50**, 997.

Wang, J. C. (1985) DNA topoisomerases. *Annu. Rev. Biochem.*, **54**, 665.

Wilson, W. D. and Jones, R. L. (1981) Intercalating drugs: DNA binding and molecular pharmacology. *Adv. Pharmacol. Chemother.*, **18**, 177.

5 Inhibitors of protein synthesis

Many antimicrobial substances inhibit protein biosynthesis. In most cases the inhibition involved one or other of the events which take place on the ribosomes. Only a few agents inhibit either amino acid activation or the attachment of the activated amino acid to the terminal adenylic acid residue of transfer RNA (tRNA). There are many chemical types to be found among the inhibitors of protein synthesis, a fact which has increased the difficulty of understanding the molecular nature of their inhibitory effects. Indeed, while the reaction which is inhibited has been identified with some precision in certain instances, the nature of the molecular interaction between the sensitive site and inhibitor remains generally elusive.

The reason lies in the complexity of the reactions leading to the formation of correctly sequenced polypeptides on the ribosome and also in the complexity of the structure of the ribosome itself. Our intention is to provide an outline of the current knowledge of the steps in protein biosynthesis. More detailed discussion is given to those specific reactions which are blocked by the inhibitors of protein biosynthesis.

RIBOSOMES

These remarkable organelles are the machines upon which polypeptides are elaborated. There are three main classes of ribosomes identified by their sedimentation coefficients. The 80S ribosomes are apparently confined to eukaryotic cells, while 70S ribosomes are found in both prokaryotic and eukaryotic cells. A unique species of 50–55S ribosome found only in mammalian mitochondria resembles bacterial ribosomes in functional organization and antibiotic sensitivity. The 80S particle dissociates reversibly into 60S and 40S subunits and the 70S into 50S and 30S subunits as the Mg^{2+} concentration of the solution is reduced. Both 80S and 70S ribosomes are composed exclusively of protein and RNA in mass ratios of approximately 50:50 and 35:65 respectively. There are three distinct species of RNA in most ribosomes, with sedimentation coefficients of 29S, 18S and 5S in 80S particles from animal cells, 25S, 16S and 5S in 80S particles from plant cells

112

and 23S, 16S and 5S in 70S particles; 55S ribosomes contain two RNA species that sediment at about 16S and 12S. In common with the 70S ribosomes of fungal mitochondria, it is doubtful whether 55S ribosomes contain 5S RNA. The protein composition of ribosomes is impressively complex. The 30S subunit of *Escherichia coli* ribosomes contains 21 proteins ('S' proteins), and the 50S subunit 34 proteins ('L' proteins). The amino acid sequences of all of these proteins are now known. In recent years great progress had been made in understanding how the ribosome is constructed. The leading groups in the field have now converged and a consensus model of the structure is available (Fig. 5.1). This reveals that the large subunit is somewhat 'armchair shaped'. The small subunit can usefully be regarded as being 'head and shoulders' shaped; in the intact ribosome it is positioned 'face' into the chair with its 'forehead' region resting on the central protuberance of the large subunit. At the time of writing the locations of 19 of the 21 'S' proteins have been mapped and rapid progress is being made in the 50S subunit.

In functional terms a translational and an exit (or secretory) domain have been defined. The translational domain is located in the head/shoulder region of the small subunit and the upper part of the armchair. The topographies of the various elongation and initiation factor binding sites are

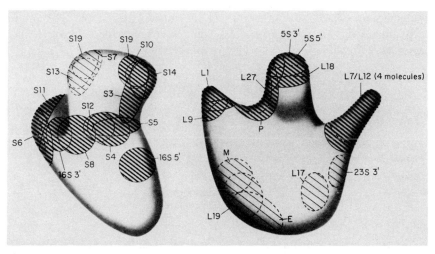

Figure 5.1 Summary map of protein, RNA and functional sites on the ribosome. Lightly shaded sites are located on the far side of the subunits. The letters P, M and E represent the peptidyltransferase site, the membrane binding site and the nascent protein exit site. The letters S and L refer to small and large subunit protein and 16S 5', 23S 3', etc. refer to the 5' and 3' ends respectively of the ribosomal RNAs. Reproduced, with permission, from *Annu. Rev. Biochem.*, **54** (1985) by Annual Reviews Inc.

well understood and represent the culmination of many years of effort by a number of groups using highly sophisticated techniques.

Recent work has also identified the exit domain's location. It is opposite the translational domain on the lower part of the large ribosomal subunit. Once these features had been defined in prokaryotic systems, very similar features were recognized in eukaryotic ribosomes. Further, eukaryotic ribosomes bind to rough endoplasmic reticulum at the exit site. This is in accord with a possible mechanism by which cells export proteins from one cellular compartment to another. As we shall see, several medically important antibiotics owe their selective antimicrobial action to a specific attack on bacterial 70S ribosomes, while the 80S particles are left unaffected. These sensitivity differences should be regarded as being due to different ribosomal domains being defined, in many cases, by several ribosomal proteins. The fact that 80S and 70S ribosomes have different proteins means that there are many domains which are unique to each system. The nature and role of the various types of ribosomal RNA are still the subjects of intensive study. There is evidence of a functional involvement of 23S RNA in the peptidyl-transferase reaction. In addition, recent proposals suggest that the codon–anticodon helix could be stabilized by co-axial stacking of one specific region of the 16S ribosomal RNA. Now that primary sequences of various ribosomal RNAs are available, progress has been made in mapping the location of the different RNA regions in the three-dimensional models of the ribosomal subunits. No doubt further details of the functional role of RNA in the ribosome structure will become apparent which may be relevant to our understanding of how some antibiotics inhibit ribosomal activity.

STAGES IN PROTEIN BIOSYNTHESIS

Formation of aminoacyl-transfer RNA

Each amino acid is converted by a specific aminoacyl-tRNA synthetase to an aminoacyladenylate which is stabilized by association with the enzyme:

$$\text{ATP + amino acid (aa)} \underset{}{\overset{\text{Enzyme}}{\rightleftharpoons}} \text{aa–AMP–Enz + PP}_i$$

Each amino acid–adenylate–enzyme complex then interacts with an amino acid-specific tRNA to form an aminoacyl-tRNA in which the aminoacyl group is linked to the 3'-OH ribosyl moiety of the 3' terminal adenosyl group of the tRNA by a highly reactive ester bond.

$$\text{aa–AMP–Enz + tRNA} \rightleftharpoons \text{aminoacyl–tRNA + AMP + Enz}$$

Initiation

The mechanism of initiation has been analysed in detail. Three protein factors, IF1, IF2 and IF3, loosely associated with the 70S ribosome are concerned with initiation. IF1 enhances the rate of ternary complex formation between mRNA, initiator tRNA and 30S ribosomal subunits, but otherwise appears to have little functional significance. IF1 has a role in promoting the dissociation of 70S ribosomes, released from previous rounds of polypeptide synthesis, into 30S and 50S subunits. Factor IF3 then binds to the 30S subunit and is also needed for the binding of natural mRNA, although not for the binding of an artificial messenger such as poly(U). The complex containing the 30S subunit, IF3 and mRNA is joined by IF2, GTP and the specific initiator tRNA, N-formylmethionyl-tRNA$_F$ (fMet-tRNA$_F$), the role of IF2 being to direct the binding of fMet-tRNA$_F$ to the specific initiator codon, AUG, site. Once it has functioned in this way, the ejection of IF2 requires the hydrolysis of one molecule of GTP to GDP and inorganic phosphate, and appears to be associated with a conformational change in the 30S subunit. The next stage involves the detachment of IF3 in the presence of a 50S subunit to permit 70S particle formation. The association of the 50S and 30S subunits is believed to involve interactions between both protein and RNA chains of the respective particles.

Initiation on 80S ribosomes is thought to resemble that on 70S ribosomes except that eukaryotic initiation *in vivo* uses unformylated Met-tRNA^{Met}. In addition, the roles of the specific eukaryotic protein factors, of which there are at least six, in initiation on 80S ribosomes, are less clearly defined than in the 70S ribosome system.

Peptide bond synthesis and chain elongation

The conventional view of polypeptide synthesis rests largely on the concept of two distinct sites on the ribosome which are usually called the acceptor (sometimes aminoacyl or A) site and the donor (sometimes peptidyl or P) site (Fig. 5.2). Recent evidence suggests that there may be an additional third tRNA-binding site, the exit (or E) site. The acceptor site is the primary decoding site where the codon of the mRNA first interacts with the anticodon region of the specific aminoacyl-tRNA. In the case of fMet-tRNA$_F$, however, binding occurs directly to the donor site. The binding of the next aminoacyl-tRNA to the acceptor site requires protein factors EFT$_s$ and EFT$_u$. The unstable factor, EFT$_u$, binds GTP and then forms a ternary complex with aminoacyl-tRNA. This complex binds to the acceptor site, with accompanying hydrolysis of one molecule of GTP. GTP hydrolysis is not essential for the binding of aminoacyl-tRNA, but in its absence the bound aminoacyl-tRNA is not available for peptide bond formation. The role of the stable factor, EFT$_s$,

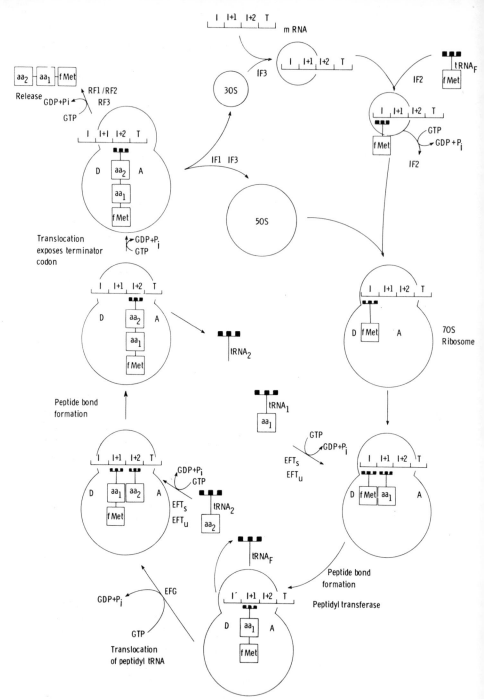

Figure 5.2 Diagrammatic scheme of major steps in polypeptide formation on a 70S ribosome. The scheme should be read clockwise starting at the top centre. I, I+1, I+2 represent initiator and successive codons and T a terminator codon on mRNA. fMet, aa_1 and aa_2 represent N-formylmethionine and two other amino acids and $tRNA_F$, $tRNA_1$ and $tRNA_2$ their specific transfer RNAs. A and D are the acceptor and donor sites. The involvement of the various protein cofactors referred to in the text is also indicated.

is to regenerate EFT_u–GTP from EFT_u–GDP by stimulating the exchange of bound GDP for a molecule of free GTP. Apparently EFT_s forms a high-affinity intermediate complex with EFT_u and GDP is lost from this intermediate.

The scene is now set for the formation of the first peptide bond. The carboxyl group of the N-formylmethionine attached to the donor site through its tRNA is 'donated' to the amino group of the adjacent amino acid at the acceptor site to form a peptide bond. The formation of the peptide bond is catalysed by peptidyltransferase which is a complex component of the 50S subunit located at the central protuberance (back of the armchair) of the 50S subunit, probably directly associated with protein L27. 5S RNA is also located in this region. The resultant dipeptide remains attached through its C-terminus to the second tRNA at the acceptor site. The dipeptidyl tRNA is EF-G (translocase) translocated from the acceptor to the donor site, still linked to the mRNA (through the codon–anticodon interaction). The deacylated tRNA may be still loosely associated with the mRNA via codon–anticodon interaction and appears to occupy the E site. This is a very loose association, which is most readily detected under suboptimal translation conditions. The third consecutive codon of the mRNA is now exposed at the acceptor site by the relative movement of the ribosome towards the 3′ end of the mRNA. The translocation step requires factor EFG and the hydrolysis of a further molecule of GTP. EFG binds to the L7/L12 region of the large subunit (one of the arms of the 'armchair'). This area has now unequivocally been implicated in GTP hydrolysis and is directly involved in EFT_u, EFG and IF2-dependent GTP hydrolysis. L7 is an N-terminus acetylated analogue of L12. This area of the ribosome is unique in that four copies of the L7/L12 complex exist per ribosome.

All of the work done so far on peptide bond formation and chain elongation on 80S ribosomes points to a general similarity to 70S ribosomes, although it is certain that there are differences of detail.

Chain termination and release

The signal for termination of the polypeptide chain is given by the appearance of one of three terminator codons, UAA, UAG or UGA, at the acceptor site. The complete polypeptide is detached from the tRNA at the C-terminal amino acid, a step that requires peptidyltransferase activity and the release factors RF1, RF2 and RF3. Factors RF1 and RF2 are concerned with the recognition of specific terminator codons, RF1 recognizing UAA and UAG and RF2, UAA and UGA. Both the binding of RF1 and RF2 to the ribosomes and their release require RF3. Cleavage of GTP is also involved in the release reaction. Release from eukaryotic ribosomes involves only one codon-recognizing release factor. There is also a requirement for GTP which is hydrolysed to GDP and inorganic phosphate during the release process. The formyl groups of the fMet ends of prokaryotic polypeptides are removed by a specific enzyme and in many proteins the methionine residue is also removed. After release of the completed polypeptide, the ribosome is liberated from the mRNA and deacylated tRNA by the action of GTP, EFG and ribosome release factor (RRF) which permits the dissociation back to 30S and 50S subunits. IF3 then binds to the 30S subunit. This prevents reassociation until a full initiation complex has once more been completed.

PUROMYCIN

The antibiotic puromycin is a unique inhibitor of protein biosynthesis, since the drug itself reacts to form a peptide with the C-terminus of the growing peptide chain on the ribosome, thus prematurely terminating the chain. This remarkable property of puromycin gave the antibiotic an important role in the elucidation of the mechanism of peptide bond formation and also of the mode of action of many other inhibitors of protein biosynthesis. We shall frequently refer to it during the discussion of the modes of action of the other inhibitors.

The structural similarity of puromycin to the terminal aminoacyladenosine moiety of tRNA was noted many years ago (Fig. 5.3) and this proved to be the key to understanding its actions. Since aminoacyladenosine is the terminal residue of tRNA in both prokaryotic and eukaryotic organisms it is

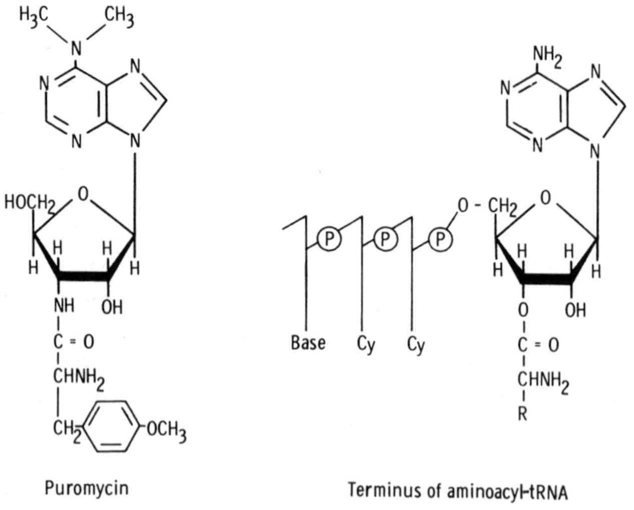

Puromycin   Terminus of aminoacyl-tRNA

Figure 5.3 Structural analogy between puromycin and the aminoacyl terminus of transfer RNA. Cy represents cytosine and R the rest of the amino acid molecule.

not surprising that puromycin terminates protein synthesis equally effectively on 70S and 80S ribosomes. The structural analogy of puromycin with
aminoacyladenosine led to the demonstration that the amino group of the
antibiotic forms a peptide bond with the acyl group of the terminal aminoacyladenosine moiety of peptidyl-tRNA attached to the ribosome. No
further peptide bond formation can occur because of the chemical stability of
the C–N bond which links the p-methoxyphenylalanine moiety of puromycin
to the nucleoside residue. Peptidyl-puromycin is released from the ribosome.

Provided that the peptidyl-tRNA is in the donor site on the ribosome, its
reaction with puromycin (the 'puromycin reaction') proceeds with no other
requirement than a normally functioning peptidyltransferase activity.
Puromycin does not, however, react with peptidyl-tRNA in the acceptor site;
factor EFG and GTP must be added in order to effect translocation of the
peptidyl-tRNA to the donor site. Only then is peptidyl-puromycin formed
and released from the ribosome. The puromycin reaction occurs fairly
readily at $0°C$, while normal chain elongation is negligible at this temperature, suggesting that puromycin has a considerable competitive advantage
over aminoacyl-tRNA in reacting with the peptidyl-tRNA. The reasons for
the greater ease of the puromycin reaction compared with normal peptide
bond synthesis are uncertain. Conceivably it could be related to the vastly
different molecular sizes of puromycin and aminoacyl-tRNA, since the bulky
aminoacyl-tRNA may have to be 'juggled' into a correct alignment at the
acceptor site in order for it to react with the peptidyl-tRNA.

A number of analogues and derivatives of puromycin have been prepared
and their ability to substitute for puromycin in the puromycin reaction
investigated. It is clear that a single benzene ring in the side chain is
necessary for activity; replacement of the p-methoxyphenylalanine with
proline, tryptophan, benzylhistidine or any aliphatic amino acid results in a
very marked loss of activity. The L-phenylalanine analogue is about half as
active as puromycin, while the D-phenylalanine analogue is completely
inactive. Replacing the p-methoxyphenylalanine residue with the S-benzyl-L-
cysteine analogue results in only a minor loss of activity which may be due to
the increased distance between the benzene ring and the free $-NH_2$ caused
by the additional S and C atoms. Since puromycin substitutes for all
aminoacyl-tRNAs equally well, the requirement for a single benzene ring in
the amino acid moiety of puromycin and its analogues is puzzling.

The aromatic ring may be involved in a hydrophobic interaction with the
terminal adenosine of peptidyl-tRNA at the donor site, thus contributing to
the formation of an intimate complex between puromycin and peptidyl-
tRNA prior to the formation of a peptide bond. In view of the structure of the
aminoacyladenosine of the tRNA terminus the requirement for linkage of the
amino acid moiety to the ribose 3′ position of puromycin is, however, not
unexpected. Puromycin substituted in the 5′ position of the ribose with

cytidylic acid is an effective peptide chain terminator, and there is an absolute requirement for cytidine in this derivative. Presumably this substitution extends the structural analogy with tRNA.

Other inhibitors of protein biosynthesis may be classified according to whether or not they inhibit the puromycin reaction. It is, however, necessary to make sure that an apparent inhibition of this reaction is not indirect. For example, inhibition of translocation results in inhibition of the puromycin reaction, since as discussed above, peptidyl-tRNA in the acceptor position cannot react with puromycin. So far it seems clear that only inhibitors which act on the 50S subunit inhibit the puromycin reaction; inhibitors acting on the 30S subunit have little or no effect.

INHIBITORS OF AMINOACYL-tRNA FORMATION

A number of naturally occurring and synthetic analogues of amino acids inhibit the formation of the aminoacyl-tRNA complex. Close analogues may become attached to the appropriate tRNA and subsequently become incorporated into abnormal proteins. Included among these are ethionine, norleucine, N-ethylglycine and 3,4-dehydroproline. Other compounds, especially several naturally occurring antibiotics such as borrelidin, furanomycin and indolmycin competitively antagonize the incorporation of the corresponding amino acids, i.e. threonine, isoleucine and tryptophan respectively, into aminoacyl-tRNA. Most of these inhibitors of aminoacyl-tRNA formation lack species specificity and hence have no useful medical application. However, indolmycin is said to be specific for prokaryotic tryptophanyl-tRNA synthetase. Chalcomycin, a macrolide antibiotic, also inhibits aminoacyl-tRNA formation by inhibiting the interaction between the activated amino acid and the tRNA.

INHIBITORS OF INITIATION-COMPLEX FORMATION
AND tRNA–RIBOSOME INTERACTION

Streptomycin

This important antibiotic is a unique member of the aminoglycoside group and has the complex chemical structure illustrated in Fig. 5.4. While there is considerable variation in the chemical structures of different aminoglycosides, they all possess a cyclohexane ring bearing basic groups in the 1 and 3 positions with oxygen substituents at 4, 5 and 6 and sometimes at 2. All these groups are in equatorial positions.

Streptomycin was discoverd by Waksman in the early 1940s and it was the first really effective drug against tuberculosis. Although it is a broad-spectrum antibiotic, active against a wide range of Gram-positive and

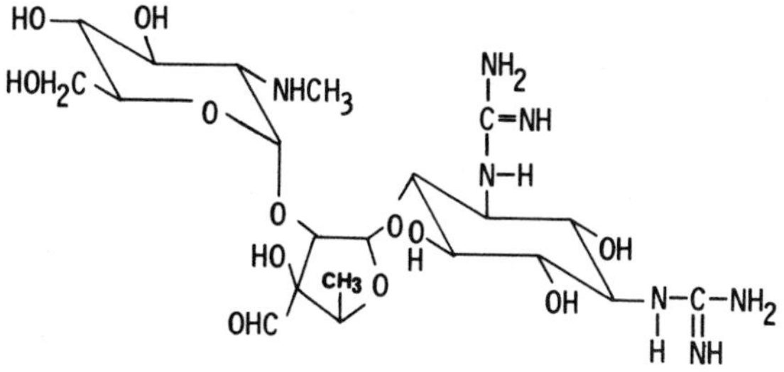

Figure 5.4 Streptomycin, the first antitubercular drug.

Gram-negative bacteria, its use has been limited by two major problems. Firstly, bacterial resistance develops readily against this antibiotic; this has been especially troublesome in the treatment of tuberculosis. Streptomycin is, therefore, usually given in combination with two other antitubercular drugs, *p*-aminosalicylic acid and isoniazid, to overcome this problem. Secondly, when the drug is given by injection (it is not absorbed from the gastrointestinal tract), it may cause irreversible deafness by damaging the 8th cranial nerve. The problem of ototoxicity is common to many of the aminoglycosides and patients must be observed closely for signs of deafness during therapy with these drugs.

Streptomycin is bactericidal, but cell death is preceded by a marked inhibition of protein biosynthesis. The inhibitory effects, studied on isolated 70S ribosomes, are complex and imperfectly understood.

(1) Streptomycin does not inhibit the initial aminoacyl-tRNA selection process, but it strongly inhibits the initiation of peptide chains. The drug also slows the elongation of partly completed chains, although even at high concentrations of streptomycin chain elongation is not completely suppressed. Peptidyltransferase activity is unaffected. These effects on initiation and elongation are attributed to a distortion of the functions of both acceptor and donor sites by streptomycin.

(2) Many studies have been carried out on the effects of streptomycin in cell-free systems using synthetic polynucleotides as messengers, and some striking effects are observed. Thus streptomycin inhibits the incorporation into peptide linkages of: (a) phenylalanine directed by poly(U), (b) histidine and threonine directed by poly(AC), (c) arginine and glutamic acid directed by poly(AG). On the other hand, streptomycin may, under some conditions, stimulate the incorporation of amino acids in the presence of synthetic messengers which do not normally code for these amino acids. For example, while streptomycin inhibits the incorporation of phenylalanine in the presence of poly(U), it stimulates the incorporation of isoleucine and serine. Again, streptomycin induces poly(C) to promote the incorporation of threonine and serine instead of proline.

In the presence of streptomycin the proof-reading selection of the correct aminoacyl-tRNA by the ribosome is markedly impaired. Misreading is not random and the following rules are more or less observed.

1. In any mRNA codon only one base is misread; this is usually a pyrimidine located at the 5' end or middle position of the codon;
2. Misreading of the base at the 3' end does not occur;
3. Misreading of purines is rare and the occurrence of these in a codon decreases the chance of misreading the codon.

Almost certainly, the induction of misreading of the genetic message by streptomycin underlies the ability of this antibiotic to suppress certain bacterial mutations.

(3) Streptomycin also promotes the messenger activity of nucleic acid molecules such as denatured DNA, ribosomal RNA and tRNA, which normally do not act as messengers in ribosomal systems.

SPECIFICITY AND SITE OF ACTION OF STREPTOMYCIN

The specific antibacterial action of this drug rests on its ability to disrupt protein synthesis on 70S ribosomes whilst having no action on 80S ribosomes. One molecule of streptomycin binds tightly ($K_d = 10^{-7}$M), but not irreversibly, to 70S ribosomes; a number of other non-specific sites ($K_d > 10^{-4}$M) also exist. The target site has been identified with some certainty by exploiting a convenient type of bacterial mutant, highly resistant to streptomycin, whose ribosomes, prepared from streptomycin-sensitive and streptomycin-resistant cells, are dissociated into 30S and 50S subunits by lowering the Mg^{2+} concentration in the medium. A ribosomal subunit 'cross-over' experiment shows that reassociated 70S particles composed of 30S subunits from resistant cells and 50S subunits from sensitive cells are resistant to streptomycin. In the opposite cross, i.e. 30S subunits from sensitive cells and 50S subunits from resistant cells, the resulting 70S ribosomes are streptomycin-sensitive. This suggests that the target site of streptomycin is on the 30S subunit, a view that is strengthened by the finding that streptomycin inhibits aminoacyl-tRNA binding not only to 70S ribosomes but also to isolated 30S subunits from sensitive cells. Using radioactive streptomycin it can be shown that the drug binds specifically to the 30S subunit but not to the 50S subunit of sensitive ribosomes. Streptomycin does not bind to the 30S subunit from resistant cells. The drug does *not* induce misreadings of mRNA translated with resistant ribosomes.

It is important in discussing the site of action of antibiotics which inhibit protein biosynthesis to consider binding domains, rather than specific interactions with one protein. Bearing this in mind, a region of the 30S subunit mapping near protein S12 appears to be the binding site for streptomycin.

Strains exhibiting intermediate resistance contain S12 proteins with specific single amino acid replacements. Thus lysine-42 may be replaced by asparagine, threonine or arginine, while lysine-87 may be replaced only by arginine. However S12 is not required for protein synthesis nor absolutely for streptomycin binding, since S12-free core particles still exhibit binding potential. The streptomycin-binding domain is located in the S3 S5 region of the subunit which is in the vicinity of the S12 binding site.

THE BACTERICIDAL ACTION OF STREPTOMYCIN

The mechanism of bacterial killing by streptomycin and other aminoglycosides has recently been re-examined by Davis (1987). His hypothesis is that misread proteins can become incorporated into the bacterial membrane, making it leaky to small molecules including the antibiotic itself. In this model the sequence of events is:

1. A small amount of antibiotic enters the cell (mechanism unclear) and causes ribosome misreading;
2. Misread protein is incorporated into the cytoplasmic membrane, creating channels which allow more antibiotic in to enhance the effect;
3. The antibiotic concentration reaches a level which blocks further protein synthesis;
4. The irreversibility of the ribosome blockade results in cell death.

It must be pointed out, however, that this remains a theory at present and that direct evidence for the role of aberrant proteins in membrane leakiness is scanty. Other explanations for the bactericidal action of aminoglycosides are possible.

Other aminoglycoside antibiotics

All aminoglycoside antibiotics have effects on protein biosynthesis, although several exert actions different from streptomycin. For example, gentamicin, which is valuable in the treatment of serious urinary tract infections, kanamycin A and neomycin C (Fig. 5.5) exhibit three separate concentration-dependent effects on isolated ribosomes. (1) At concentrations below $2\,\mu g\,ml^{-1}$ there is strong inhibition of total protein synthesis, but little induction of mRNA misreading. (2) Between 5 and 50 $\mu g\,ml^{-1}$, misreading becomes evident, especially of reading through the termination signals. Protein synthesis may therefore increase through the accumulation of abnormally long polypeptides as the ribosomes continue past the end of one message and on to the next. (3) Higher concentrations re-establish inhibition of protein synthesis. Recent studies using radiolabelled kanamycin A showed that each ribosomal subunit has one strong binding site for this agent as well as a number of weak binding sites. Gentamicin and neomycin compete with the kanamycins suggesting similar binding domains which must be distinct from the streptomycin-binding area since no such competition is observed with streptomycin. Recent data suggest that several aminoglycosides exert their bactericidal effects (as distinct from a bacteriostatic action), by binding to the 30S subunit, possibly by directly interacting with 16S RNA. Further research will be necessary to define clearly the basic mechanism of action of the aminoglycosides on ribosomal function.

Spectinomycin is usually included in the aminoglycoside group, even though it lacks an amino sugar residue (Fig. 5.5). Unlike the previously mentioned aminoglycoside antibiotics, it is bacteriostatic rather than bactericidal in action. Its effects on protein synthesis are also markedly different from those of the other aminoglycosides. While it inhibits protein biosynthesis in bacterial cells and in cell-free systems containing 70S ribosomes,

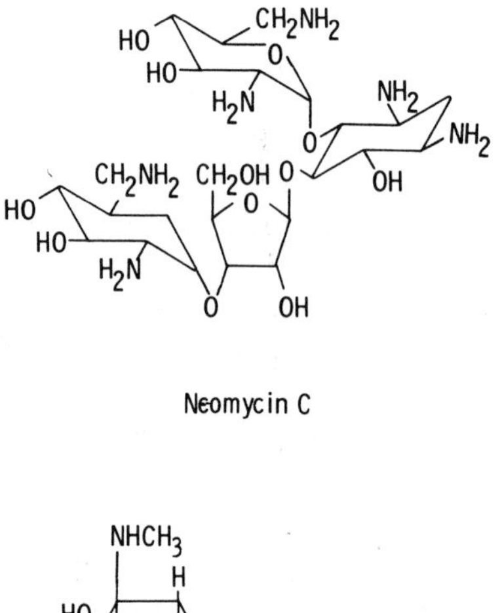

Neomycin C

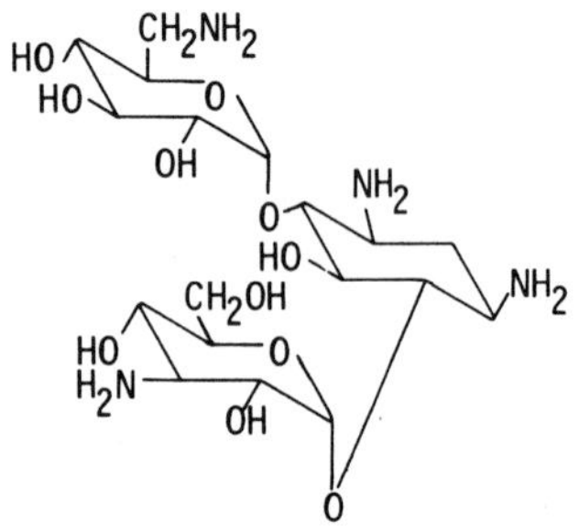

Kanamycin A

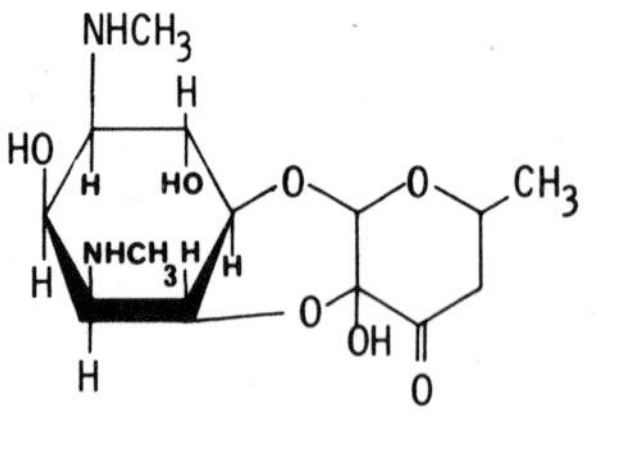

Spectinomycin

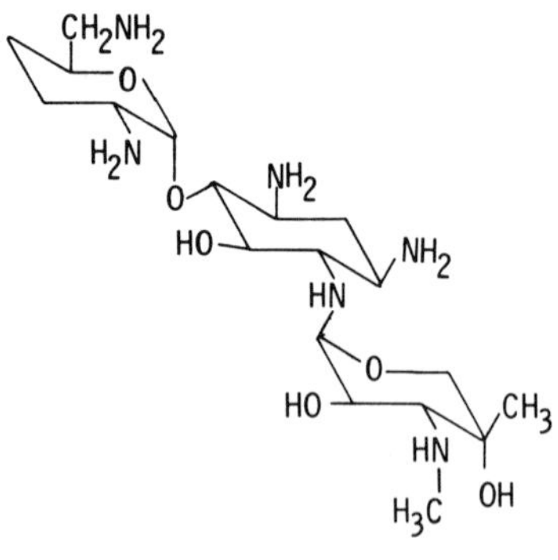

Gentamicin C_{1a}

Figure 5.5 Other aminoglycoside antibiotics. Spectinomycin is more appropriately described as an aminocyclitol antibiotic as it contains an inositol ring with two of its OH groups substituted by methylamino groups.

spectinomycin does not induce ambiguity in the readout of mRNA. Spectinomycin may inhibit an initial translocation step as it has no effect on codon recognition, peptide bond formation, or chain termination and release. With the emergence of β-lactamase-producing *Neisseria gonorrhoeae*, spectinomycin is finding extensive clinical use as the agent of choice for these infections.

Tetracyclines

Four important members of this group are illustrated in Fig. 5.6. The tetracyclines are broad-spectrum antibiotics and they are also effective against rickettsial organisms, mycoplasmas and certain protozoa. The antibiotic (bacteriostatic) activity of the tetracyclines depends on their direct inhibition of protein biosynthesis. Unlike the aminoglycoside antibiotics, the tetracyclines inhibit protein biosynthesis on both 70S and 80S ribosomes, although 70S ribosomes are rather more sensitive. However, the tetracyclines are much more effective against protein synthesis in intact bacteria than against eukaryotic cells because of selective absorption by bacterial cells (Chapter 7).

Studies of the effects of the tetracyclines on the tRNA–ribosome interaction show that they inhibit the binding of aminoacyl-tRNA to the acceptor site on the ribosome but have little effect on binding to the donor site except at high drug concentrations. The binding of fMet-tRNA$_F$ to the ribosome is about one-tenth as sensitive to tetracycline as the binding of other aminoacyl-tRNAs, since fMet-tRNA$_F$ binds to the donor site rather than to the acceptor site. The tetracyclines do not directly inhibit formation of the

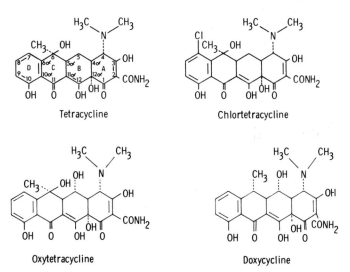

Tetracycline

Chlortetracycline

Oxytetracycline

Doxycycline

Figure 5.6 Four major tetracycline antibiotics.

peptide bond or the translocation step except at high concentrations. They have no effect on the hydrolysis of GTP to GDP that is required for functional binding of aminoacyl-tRNA to the acceptor site. Possibly the tetracyclines uncouple the hydrolysis of GTP from the binding reaction. The tetracyclines also inhibit peptide chain termination and release but this is assumed to be related to the fact that the release factors normally bind to the termination codons in the A site; this is prevented by tetracycline, in a manner similar to the inhibition of aminoacyl-tRNA binding. It is doubtful that these effects contribute to the antibacterial action of these agents.

The nature of the interaction between tetracycline and the ribosome has been the subject of numerous studies and even more speculation. The consensus view is that there is one strong binding site per 70S ribosome ($K_d = 1–30\,\mu\text{M}$). There are also several hundred weaker binding sites ($K_d = 60\text{mM}$). In order to define more precisely the location of the strong binding site, careful photoaffinity labelling studies were performed. These revealed extensive labelling of protein S7. This evidence, together with the recent topological data which reveal that S7 is located near the contact region between the two subunits, may explain why the 50S subunit has also been implicated in the ribosome's interaction with tetracycline.

Once again, the molecular explanation of the inhibitory effects of the tetracycline eludes us. The relationship between chemical structure and antibacterial activity has been worked out in some detail. There is, however, a specific permeation mechanism for the entry of tetracyclines into bacterial cells (see Chapter 7) that has its own structural requirements. It is possible, therefore, that, although some analogues of the tetracyclines may be active inhibitors of protein synthesis, they cannot penetrate into the cell and so lack antibacterial activity.

The limited investigations on the structural requirements of tetracyclines for the inhibition of protein synthesis on isolated ribosomes indicate a number of modifications in structure that significantly affect the degree of inhibition.

1. Chlorination of the 7 position significantly increases inhibitory activity;
2. Epimerization of the 4-dimethylamino group significantly decreases activity (Fig. 5.7);
3. Both 4α,12α-anhydro- and 5α,6-anhydro-tetracyclines (Fig. 5.7) are much less active than tetracycline.
4. Ring opening of chlortetracycline and tetracycline to give the iso derivatives (Fig. 5.7) and of oxytetracycline to give α and β isomers of apo-oxytetracycline (Fig. 5.7) results in loss of activity.
5. Replacement of the amidic function at C_2 with $C\equiv N$ to give tetracycline nitrile results in a marked loss of potency compared with tetracycline.

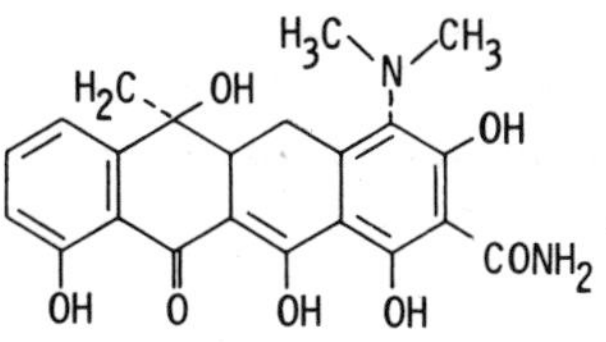

4α, 12α -Anhydrotetracycline

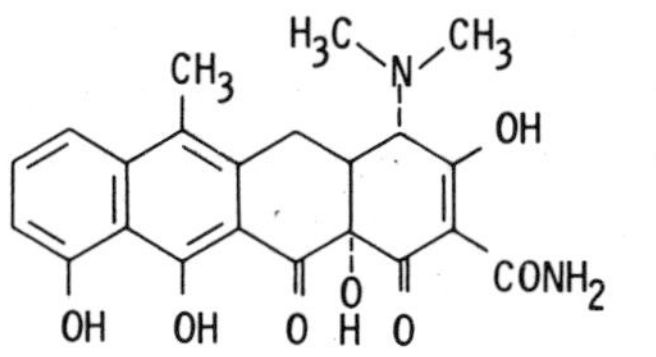

5a, 6-Anhydrotetracycline

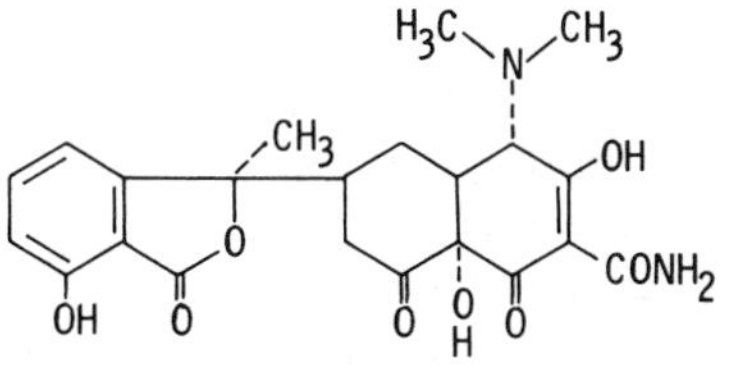

Isotetracycline

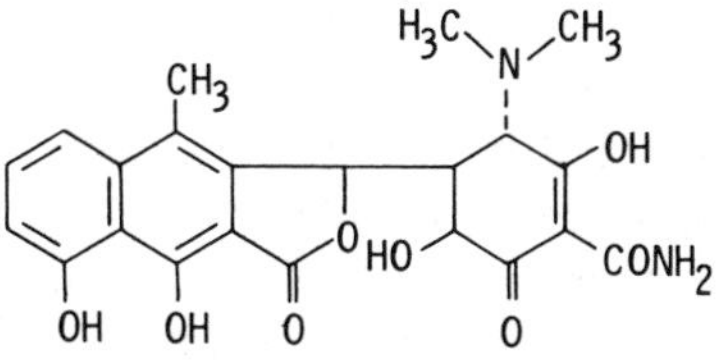

Apo-oxytetracycline

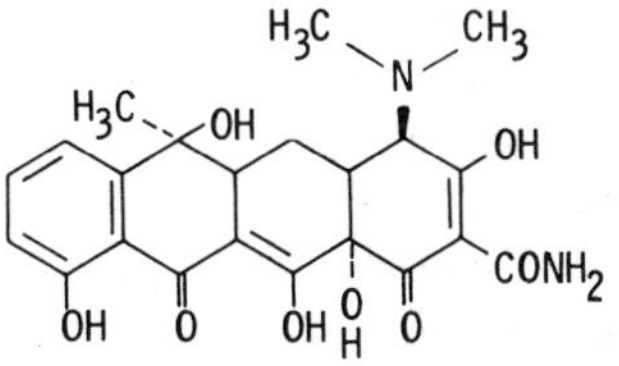

4-Epitetracycline

Figure 5.7 Tetracycline derivatives with greatly reduced antibiotic activity.

The ability of the tetracyclines to chelate polyvalent cations may have a bearing on their inhibition of protein biosynthesis. It is tempting to propose that Mg^{2+} attached to the phosphate groups of the ribosomal RNA may act as a link between the ribosome and tetracycline molecules. The concentration of free Mg^{2+} in the vicinity of the ribosomes in the cell will, however, have a major influence on the ability of tetracyclines to interact with ribosomal Mg^{2+}. The 11,12β-diketone system, the 12α- and the 3-hydroxyl groups have all been implicated as possible complexing sites for polyvalent cations. An alternative suggestion based on circular dichroism studies on chlortetracycline complexes with Ca^{2+} and Mg^{2+} is that chelation requires the bending of ring A back towards ring B and C so that the oxygen atoms at positions 11 and 12 together with those at positions 2 (amide 0) and 3 form a co-ordination site into which the metal atom fits. The structural modifications to the tetracycline molecule described above, which affect its ability to inhibit protein synthesis on the ribosomes, may also affect its metal-complexing properties, but no final conclusion as to the nature of the molecular interaction between the tetracyclines and ribosome can be drawn at present.

INHIBITORS OF PEPTIDE BOND FORMATION AND TRANSLOCATION

Chloramphenicol

This antibiotic (Fig. 5.8) has an antimicrobial spectrum similar to that of the tetracyclines. However, its most important medical uses are in the treatment of typhoid fever and bacterial meningitis because serious side effects have limited the use of chloramphenicol as a general antibacterial agent. Although originally a natural product, the relative chemical simplicity of chloramphenicol has made its production by chemical synthesis commercially more attractive than the fermentation route.

The bacteriostatic action of chloramphenicol is due to a specific primary inhibition of protein biosynthesis on 70S ribosomes; it is completely inactive against 80S ribosomes.

Studies with radioactively labelled chloramphenicol show that it binds exclusively to the 50S subunit to a maximum extent of one molecule per subunit. The binding is completely reversible. Structurally unrelated antibiotics such as erythromycin and lincomycin, which also interfere with the function of the 50S subunit, compete with chloramphenicol for the binding region. The aminoglycosides have no effect on the binding of chloramphenicol to the 50S particle. Tetracycline, whose inhibitory action is on the 30S subunit and yet binds to both subunits also fails to affect chloramphenicol binding.

Much effort has gone into identifying the proteins that may be involved in the interaction of chloramphenicol with the 50S particle. The failure to isolate a resistant mutant whose ribosomes contain altered proteins has, so far, precluded the direct identification of a chloramphenicol target protein. As discussed later, there is evidence of RNA base modification being a mechanism by which ribosomes can become resistant to this drug.

Recent studies, using the technique of affinity immune electron micros-

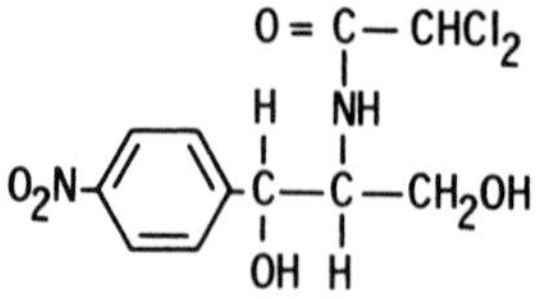

Figure 5.8 Chloramphenicol. The active form is the D-*threo* isomer.

copy, have clearly indicated that the peptidyltransferase area of the ribosome containing the proteins L15, L18 and L27 provides the binding region for chloramphenicol. This region has similarly been identified as the main puromycin site. There is also good evidence for S14 being involved in this area as well. Because of the clear functional association with the peptidyltransferase activity, this region should correspond to the location of the 3' end of tRNAs both in the ribosomal A site and P site. This scheme is supported by following observations. (1) Chloramphenicol inhibits the puromycin-dependent release of nascent peptides from 70S ribosomes. (2) Chloramphenicol interferes with a model peptide-forming reaction catalysed by 50S ribosomal subunits. This preparation contains N-formylmethionyl-ACC-AAC (the terminal hexanucleotide fragment of fMet-tRNA$_F$), 50S subunits and puromycin. In the presence of alcohol the dipeptide analogue N-formylmethionylpuromycin is synthesized and released.

The present view is that chloramphenicol inhibits peptidyltransferase activity by binding to a region on the 50S subunit close to the site that binds aminoacyl-tRNA in the peptidyltransferase centre. Since the interaction of chloramphenicol with 70S ribosomes is rapidly reversible, bacterial cells previously inhibited by chloramphenicol soon resume protein synthesis and growth when they are transferred to drug-free medium. Direct binding, or association, of a number of antibiotics, including chloramphenicol to ribosomal RNA has recently been proposed and will be discussed more fully in the next section.

Erythromycin

This complex antibiotic (Fig. 5.9) is a member of the macrolide group which is characterized by molecular structures containing large lactone rings linked with amino sugars through glycoside bonds. Erythromycin is a medium-spectrum antibacterial agent, effective against many Gram-positive bacteria but only against relatively few Gram-negative organisms. The specific antibacterial action of erythromycin depends on its blockade of protein synthesis on 70S ribosomes while having no action on 80S ribosomes. Like chloramphenicol, erythromycin binds exclusively to the 50S subunit. Although it antagonizes the binding of chloramphenicol, erythromycin does not inhibit peptide bond formation or translocation.

Uniquely amongst the antibiotics so far discussed, erythromycin binds specifically to isolated L15 ribosomal protein. As mentioned above, this protein is part of the peptidyltransferase region of the 50S subunit. A pattern of collateral resistance to macrolides, lincomycin and streptogramin B correlates with the N_6 dimethylation of residue 2058 (adenine) of 23S RNA. Specifically erythromycin resistance is associated with point mutations at residues 2058 or 2611. Residues 2447, 2451, 2452, 2503 and 2504 are also

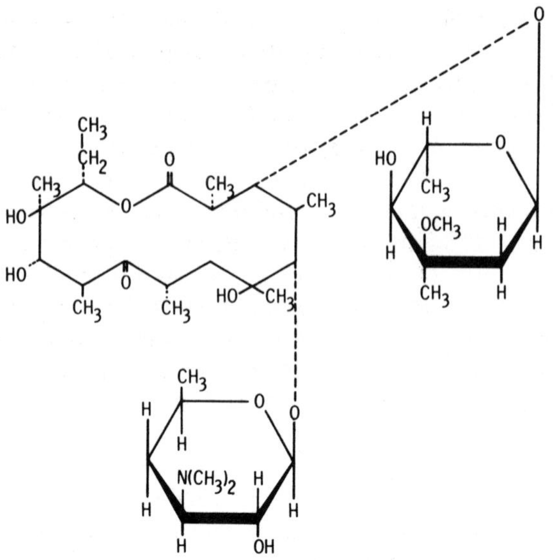

Figure 5.9 Erythromycin, a macrolide antibiotic.

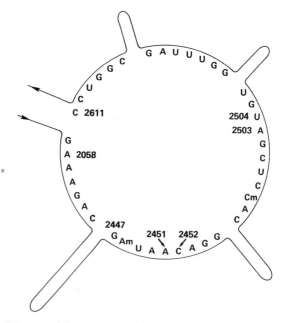

Figure 5.10 Diagram of the structure of the 23S RNA region involved in the peptidyl-transferase centre of the *Escherichia coli* ribosome.

involved in chloramphenicol resistance. All this information comes together in a fascinating manner when the folding pattern of this RNA sequence is examined (Fig. 5.10). This indicates that these two antibiotics bind in a mutually exclusive manner and strongly suggests that this RNA sequence must be located in the peptidyltransferase region of the 50S ribosomal subunit.

The mechanism of action of erythromycin and of macrolides in general has been the source of much confusion and many apparently contradictory results. Recent progress has greatly clarified these difficulties and a general mechanism of action can be suggested. It is proposed that macrolide antibiotics stimulate the dissociation of peptidyl-tRNA from ribosomes, possibly during an attempted translocation step thereby interrupting the completion of peptide chains.

Lincomycin and clindamycin

Lincomycin and its clinically more useful analogue, clindamycin, (Fig. 5.11) are active against many Gram-positive bacteria but are inactive against Gram-negative species. Although they inhibit protein synthesis on ribosomes from Gram-positive bacteria such as *Bacillus stearothermophilus*, ribosomes from Gram-negative bacteria are generally resistant. This unusual differential sensitivity of ribosomes apparently results from the inability to bind effectively to ribosomes from Gram-negative bacteria. 80S ribosomes from eukaryotic cells are also resistant to lincomycin and clindamycin.

The 50S subunit bears the target site for lincomycin and the antibiotic binds exclusively to this subunit. The involvement of the residue A2058 in lincomycin resistance, and perhaps binding, has already been discussed above. It is therefore likely that the inhibitory action of lincomycin and clindamycin is similar to that of the macrolides.

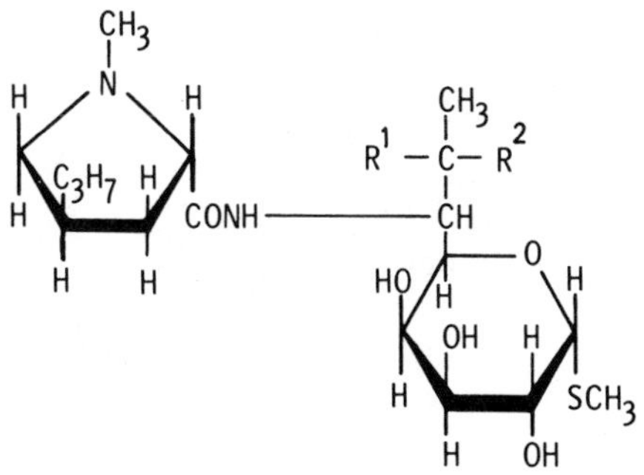

Figure 5.11 Lincomycin (R^1=OH, R^2=H) and clindamycin (R^1=H, R^2=Cl). Clindamycin is chemically manufactured from the natural product lincomycin.

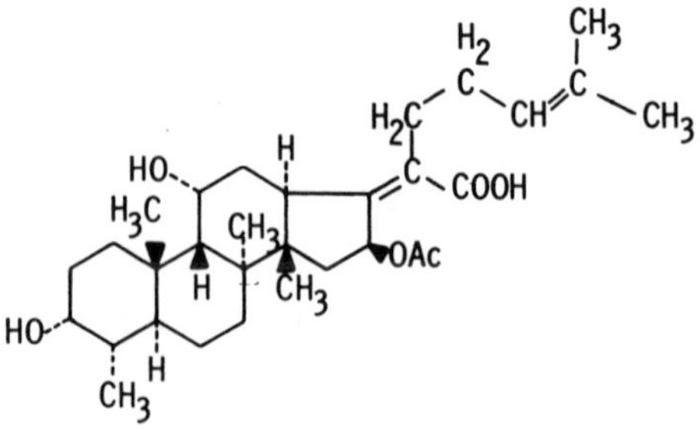

Figure 5.12 Fusidic acid, an antibiotic with a steroid-like structure.

Fusidic acid

Fusidic acid belongs to a group of steroidal antibiotics (Fig. 5.12). It inhibits the growth of Gram-positive but not Gram-negative bacteria and is used clinically to treat Gram-positive infections that are resistant to more widely used agents. Its lack of activity against Gram-negative bacteria may be due to a failure to gain access to their ribosomes since it inhibits protein synthesis on ribosomes from Gram-negative and Gram-positive species *in vitro*. Fusidic acid also blocks protein synthesis in cell-free preparations from yeast and reticulocytes, both of which, of course, contain 80S ribosomes.

Addition of fusidic acid to 70S ribosomes *in vitro* prevents the translocation of peptidyl-tRNA from the acceptor to the donor site and also inhibits the EFG-dependent cleavage of GTP to GDP. The inhibition is overcome by the addition of excess EFG. Bacteria resistant to fusidic acid contain EFG that is also resistant to the drug, and translocation proceeds normally in its presence. All this points to factor EFG as the target protein for fusidic acid. However, fusidic acid does not directly inhibit GTP hydrolysis. Indeed, in the presence of the drug, an initial burst of GTP hydrolysis can be observed which then slows to zero. Fusidic acid forms a stable complex with EFG, GDP and the ribosome which is unable to release EFG for a further round of translocation and GTP hydrolysis. It appears that fusidic acid works in essentially the same way on 80S ribosomes by stabilizing the EF2–GDP–ribosome complex (EF2 corresponds to the prokaryotic EFG). The lack of toxicity of fusidic acid against mammalian cells is probably because the drug does not achieve an intracellular concentration capable of inhibiting the ribosomes.

Thiostrepton

Thiostrepton is a complex bicyclic modified peptide antibiotic (Fig. 5.13), which binds with very high affinity ($K_D < 1$nM) to a site on the 50S ribosomal subunit involved in GTP-dependent reactions. It has been shown that the

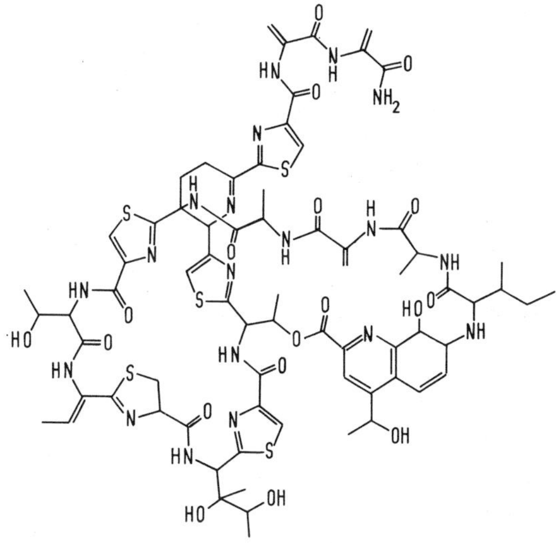

Figure 5.13 Thiostrepton, a remarkably complex inhibitor of ribosomal function.

antibiotic binds to 23S RNA alone with a K_D of $1\,\mu\mathrm{M}$ but the addition of protein L11 (which shows no affinity itself for the drug) lowers the dissociation constant by three orders of magnitude. This protein L11 is a close neighbour of L10 and the L7/L12 dimer complex which is located at the GTPase centre of the ribosome. The thiostrepton-producing organism, *Streptomyces azureus*, produces a specific methylase which acts on 23S RNA to modify residue number A1067, which is part of the oligonucleotide region protected by the presence of L11 from ribonuclease attack. The formation of $2'$-O-methyladenosine at this position renders the ribosomes totally resistant to the drug. This is an elegant correlation between a unique base modification and drug resistance and also once more reminds us of the vital and dynamic role RNA plays in the ribosomal structure.

Cycloheximide

Referred to in older papers as actidione, cycloheximide (Fig. 5.14) is unusual as it specifically inhibits 80S ribosomes but has no action on 70S ribosomes.

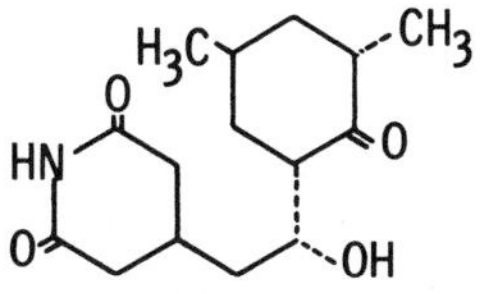

Figure 5.14 Cycloheximide, a specific inhibitor of 80S ribosomes.

It is toxic to a wide range of eukaryotic cells including protozoa, yeasts, fungi and mammalian cells. This lack of selectivity precludes the clinical use of cycloheximide.

Mostly the antibiotic is used as an experimental tool for the inhibition of protein synthesis in eukaryotic cells, and to exclude fungi from bacterial cultures.

There are considerable variations in the sensitivity of 80S ribosomes from different species to cycloheximide; for example, ribosomes from *Saccharomyces cerevisiae* are strongly inhibited to cycloheximide, while those from *Saccharomyces fragilis* are resistant. These differences in ribosomal sensitivity have been used to locate the subunit specificity of cycloheximide. Cross-over experiments with the 60S and 40S subunits from *Saccharomyces cerevisiae* and *Saccharomyces fragilis* show that sensitivity to cycloheximide resides in the 60S subunit. Intrinsically resistant ribosomes apparently have modified 60S subunits which are unaffected by the antibiotic.

Cycloheximide interferes with the translocation of peptidyl-tRNA from the acceptor to the donor site on the ribosome and, as with other compounds having this kind of action, the point of attack is on the larger of the two ribosomal subunits. Opinion is still divided as to whether the hydrolysis of GTP associated with translocation is inhibited by cycloheximide. The likelihood is that cycloheximide directly inhibits the function of the eukaryotic translocation factor EF2.

NUCLEIC ACID SYNTHESIS DURING INHIBITION OF PROTEIN SYNTHESIS

Those inhibitors of protein biosynthesis that do not cause cell death allow the synthesis of RNA and DNA to continue for at least a limited period. In some cases the continued synthesis of nucleic acids provides useful indirect evidence for a selective action of a drug on protein synthesis. For example, the inhibition of protein biosynthesis by the tetracycline antibiotics was formerly regarded as secondary to their interference with energy-yielding reactions in the cell. The demonstration that nucleic acid synthesis persisted after inhibition of protein synthesis by tetracyclines proved that the latter effect could not be due to inhibition of energy production, since nucleic acid synthesis also requires biological energy. The continuation of nucleic acid synthesis also indicates that a drug does not inhibit protein synthesis by interfering with the synthesis of mRNA.

Originally it was thought that the inhibition of protein synthesis had little or no effect on DNA synthesis, since the latter process continues for some considerable time after the addition of inhibitors of protein synthesis. It is now realized, however, that the inhibition of protein synthesis allows only the completion of a previously initiated cycle of replication of the bacterial chromosome. After the completion of this cycle, DNA synthesis ceases. The continued biosynthesis of proteins associated with the initiation of DNA synthesis is essential for a fresh round of DNA replication.

EFFECTS OF INHIBITORS OF 70S RIBOSOMES ON EUKARYOTIC CELLS

Subcellular organelles in eukaryotic cells, such as mitochondria or chloroplasts in photosynthetic organisms, contain ribosomes that more closely resemble 70S than 80S ribosomes both in size and sensitivity to ribosomal inhibitors. Although only a small proportion of the protein of the subcellular organelles is synthesized on the organelle-specific ribosomes, the inhibition of these ribosomes may underlie the action of some antibiotics on eukaryotic cells. For example, yeast cells grown in the presence of chloramphenicol, erythromycin or lincomycin become deficient in various cytochromes through interference with the biogenesis of the mitochondria. Streptomycin and erythromycin cause the loss of the photosynthetic organelles from *Chlamydomonas* and *Euglena* and this may be associated with antibiotic attack on the chloroplast ribosomes. Despite the excellent antibacterial activity of chloramphenicol its clinical use is severely limited by a tendency to damage the bone marrow. A concentration of 25–30 μg of chloramphenicol ml^{-1} of blood maintained for 1–2 weeks leads to an accumulation of nucleated erythrocytes in the marrow, indicating an interference with the normal maturation process. Normal erythropoiesis usually resumes after withdrawal of the drug, but very occasionally, i.e. not more than 1 in 20 000 cases, a more serious defect develops in the marrow which leads irreversibly to the loss of both white and red cell precursors. The biochemical basis for the marrow toxicity of chloramphenicol has not been established. However, an action of the drug on mitochondrial ribosomes, with a consequent loss of respiratory function in the marrow cells, cannot be ruled out. The very low incidence of the fatal irreversible form of chloramphenicol toxicity indicates a special sensitivity in those few individuals who succumb to it.

FURTHER READING

Chopra, I. (1985) Mode of action of the tetracyclines and the nature of bacterial resistance to them. *Handb. Exp. Pharmacol.*, **78**, 313.

Cundliffe, E. (1986) Thiostrepton and the ribosomal GTPase center. In *Structure, Function and Genetics of Ribosomes* (eds B. Hardesty and G. Kramer), Springer Verlag, p. 590.

Davis, B. D. (1987) Mechanism of bactericidal action of aminoglycosides. *Microbiol. Rev.*, **51**, 341.

Gale, E. F., Cundliffe, E., Reynolds, P. E., Richmond, M. and Waring, M. J. (1981) *The Molecular Basis of Antibiotic action*, 2nd edn., John Wiley and Sons, London.

Hardesty, B. and Kramer, G. (eds) (1986) *Structure, Function and Genetics of Ribosomes*, Springer Series in Molecular Biology, Springer Verlag.

Jelenc, P. C. and Kurland, G. G. (1984) Multiple effects of kanamycin on translational accuracy. *Mol. Gen. Genet.*, **194**, 195.

Kaji, A. and Ryoji, M. (1979) Tetracycline. In *Antibiotics*, Vol. 5, Part 1 (ed. F. E. Hahn), Springer Verlag, p. 304.

Lake, J. A. (1985) Evolving ribosome structure: Domains in Archaebacteria, Eubacteria, Eocytes and Eukaryotes. *Annu. Rev. Biochem.*, **54**, 507.

Menninger, J. R. and Otto, D. P. (1982) Erythromycin, carbomycin and spiramycin inhibit protein synthesis by stimulating the dissociation of peptidyl-tRNA from ribosomes. *Antimicrob. Ag. Chemother.*, **21**, 811.

Omura, S. (1984) *Macrolide Antibiotics: Chemistry, Biology and Practice*, Academic Press, London.

Skeggs, P. A., Thompson, J. and Cundliffe, E. (1985) Methylation of 16S rRNA and resistance to aminoglycoside antibiotics in clones of *Streptomyces lividans* carrying DNA from *Streptomyces tenjimariensis*. *Mol. Gen. Genet.*, **200**, 415.

Stoffler, G. and Stoffler-Meilicke, M. (1981) Structural organisation of *E. coli* ribosomes and localisation of functional domains. In *International Cell Biology* (ed. H. G. Schweiger), Springer Verlag, p. 93.

Stoffler, G. and Stoffler-Meilicke, M. (1984) Immuno-electron microscopy of ribosomes. *Annu. Rev. Biophys. Bioeng.*, **13**, 303.

6 Antifungal, antiprotozoal and antiviral agents

In addition to bacterial disease many infections are caused by fungi, viruses and protozoa. Chemotherapy of these infections has been somewhat neglected because of the difficulty in designing safe and selective agents, but in recent years chemical ingenuity has led to some major breakthroughs.

COMPOUNDS USED TO TREAT PROTOZOAL DISEASES

Protozoal disease is a serious problem in man and animals, especially in the hotter areas of the world. Chemotherapy of such diseases has a long history which predates antibacteral therapy by several decades. Many drugs have been developed for use in malaria, trypanosomiasis, amoebiasis and other protozoal diseases, but with few exceptions their mode of action is uncertain or obscure. This is largely due to the great technical difficulty of culturing protozoa under controlled experimental conditions, compared with the simplicity of growing bacteria and fungi. Some compounds developed as antibacterials or antifungal agents are effective against protozoa in man. More commonly the antiprotozoal drugs stand on their own and possess little useful antibacterial activity.

One major problem in the search for antiprotozoal agents is the common occurrence of gross differences in morphology in various stages of the life cycle of protozoa. Usually these forms exist in different environments which require the organism to employ alternative metabolic processes. The complex changes which occur indicate that the protozoan can adapt its metabolism and therefore may be able to circumvent inhibition by potential drugs.

In addition to these problems there is the question of similarity of protozoan biochemistry to that of mammalian cells. Selectivity of drugs between host and parasite therefore depends upon differences in penetration, the use of alternative metabolic pathways, the relative importance of those pathways and finally differences in the affinities of similar enzymes in pathogen and host for drug molecules. The antiprotozoal agents will be considered in groups with similar types of targets.

137

Energy metabolism

Many antiprotozoal agents act on the energy metabolism of the organism. Most parasitic protozoa use glucose as a major energy source and use the Embden–Meyerhof pathway to produce pyruvate. Beyond this step the metabolic routes are species specific.

Arsenic and antimony compounds still retain a place in the treatment of protozoal infections. Like arsenite, the trivalent organic arsenicals react readily with thiol compounds and especially dithiols such as lipoic acid. Consequently they inhibit enzyme systems which catalyse the oxidative decarboxylation of 2-oxo acids, in which lipoic acid acts as a cofactor. These, however, are unselective actions and preferential toxicity to trypanosomes is probably explained by selective uptake. Experiments with *Trypanosoma rhodesiense* indicate a marked increase in phosphoenolpyruvate following treatment with $2 \mu M$ melarsen (a trypanocidal arsenical) (Fig. 6.1). This is attributed to inhibition of phosphokinase, though the isolated enzyme is only inhibited at much higher concentrations. Muscle pyruvate kinase is relatively insensitive to the arsenical. Since the trypanosome relies on glycolysis for its energy production, inhibition of pyruvate kinase deprives it of its source of ATP. Similarly, organic antimonials inhibit phosphofructokinase from schistosomes at one-hundredth of the concentration required to inhibit the corresponding mammalian enzyme.

The trypanocidal drug suramin (Fig. 1.2) disrupts the function of the glycerophosphate pathway by inhibiting both glycerol 3-phosphate oxidase and NAD^+-dependent glycerol 3-phosphate dehydrogenase. These inhibitory actions interfere with the reoxidation of NADH and lead to a shift towards the production of glycerol and a reduction in ATP production, which is lethal for the trypanosome.

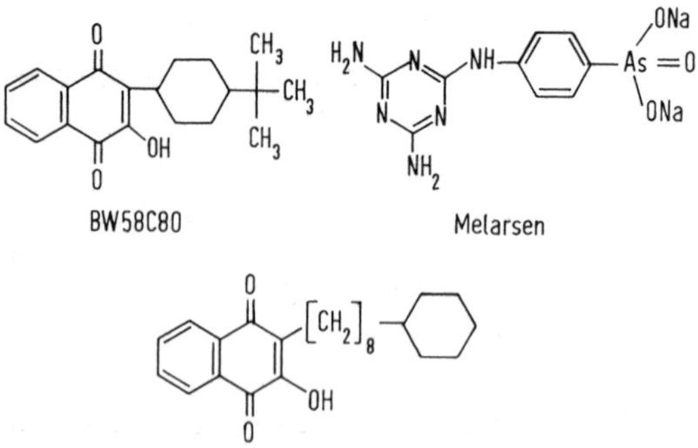

Figure 6.1 Compounds inhibiting protozoa by interfering with their energy metabolism.

The mitochondria of several blood-borne parasites are affected by hydroxynaphthoquinones such as BW58C80 and menoctone (Fig. 6.1). The parasites use ubiquinone-8 rather than ubiquinone-10 which is produced by mammalian cells, and the selectivity of these naphthoquinones is due to competition with the ubiquinone-8.

Cell membranes

The cell membranes of protozoa may represent another useful target for toxic agents. Amphotericin B (Fig. 6.6) is active against intracellular amastigotes of some *Leishmania* species, but is little used because of its overt host toxicity. Its activity probably results from the existence of ergosterol rather than cholesterol in the cell membranes (see later).

A number of antibacterial ionophoric antibiotics are highly active against coccidia, e.g. monensin (Fig. 3.7), but it is unclear whether their anticoccidial action depends upon their ionophoric activity in cell membranes. The antifungal drug miconazole which inhibits sterol biosynthesis in fungi (see later) is also active against the protozoan parasite *Leishmania*. However its antiprotozoal action is not thought to be via sterol biosynthesis but is in some way linked to the oxygen status of the environment, as its effects are more marked when the oxygen tension is high.

The widely used antimalarial chloroquine (Fig. 4.11) and related compounds such as mepacrine (Fig. 1.2) are believed to kill the intraerythrocytic parasites by an action on the protozoal membranes. It has long been recognized that these compounds are concentrated in parasite membranes. Trophozoites of *Plasmodium*, when in erythrocytes, digest the haemoglobin to obtain amino acids and in the process form a haemolytic porphyrin, haematin. The parasite normally produces a substance that binds to haematin and converts it to the inert malaria pigment haemazoin which protects the parasite and host membranes from the lytic action of haematin. Chloroquine forms a multimeric complex with haematin, which subsequently dissociates to release haematin into the membranes which are then disrupted.

Biosynthesis of essential cofactors

Interference with the uptake or biosynthesis of nutrients and metabolites can restrict the availability of essential cofactors and is responsible for the activity of some antiprotozoal drugs.

Like many bacteria, several parasites cannot use preformed folates and must synthesize them. The parasites causing coccidiosis, malaria and toxoplasmosis, with few exceptions, depend on the folate precursor *p*-aminobenzoate and are therefore susceptible to inhibition by sulphonamides, sulphones and related drugs which competitively antagonize *p*-

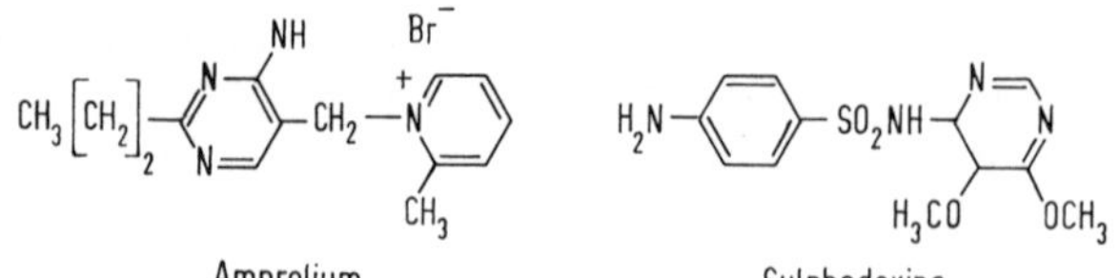

Amprolium   Sulphadoxine

Figure 6.2 Antiprotozoal agents that interfere with the biosynthesis of essential cofactors.

aminobenzoate. Further along the folate pathway, the conversion of dihydrofolate to tetrahydrofolate by dihydrofolate reductase is susceptible to inhibition (see Chapter 4). The affinity of this enzyme in the asexual intraerythrocyte stage of malarial parasites for pyrimethamine (Fig. 4.3) is 2000 times greater than the affinity of the mammalian enzyme. As a result pyrimethamine is an effective anti-malarial drug.

Marked potentiation is obtained when protozoa are exposed to an appropriate combination of pyrimethamine and a sulphonamide such as sulphadoxine (Fig. 6.2). This combination is similar to that of the antibacterial drug cotrimoxazole which combines an inhibitor of dihydrofolate reductase, trimethoprim, with sulphamethoxazole (Chapter 4). Amprolium (Fig. 6.2), a widely used coccidiostat, is structurally similar to thiamine. It is thought to act at low concentrations by blocking the access of thiamine to coccidia in the gut. Higher concentrations also prevent uptake of thiamine by the host.

Macromolecular synthesis

Protein synthesis in protozoa is essentially similar to that in mammals and the potential for finding useful drugs that act on the process is relatively limited. However, two types of antibacterial agents, which inhibit protein synthesis, surprisingly show activity against certain protozoa. Several tetracyclines and lincomycin (Chapter 5) have been used in the treatment of *Plasmodium falciparum* malaria. The basis of their selective antiprotozoal activity is however uncertain.

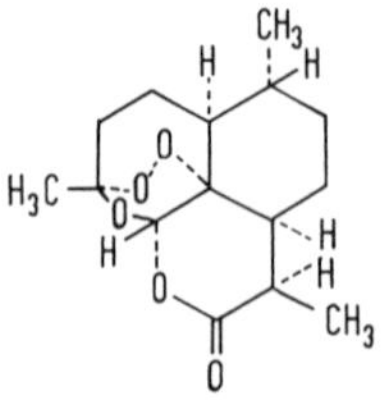

Figure 6.3 The sesquiterpene antiprotozoal agent qinghaosu.

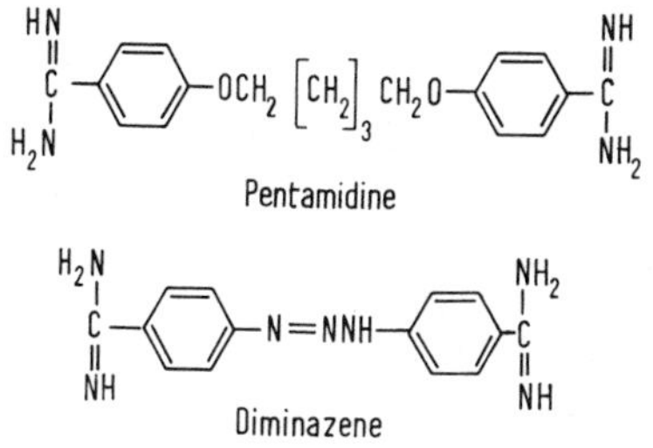

Pentamidine

Diminazene

Figure 6.4 Compounds interacting with protozoan DNA.

The natural product qinghaosu (Fig. 6.3) also known as artemisinine, was extracted by Chinese workers from a traditional plant recipe. This sesquiterpene lactone antimalarial inhibits protein synthesis in cultures of *Plasmodium falciparum*. Later on nucleic acid synthesis is also affected but this is probably a secondary effect. An important factor in the antiparasitic action of qinghaosu is the selective uptake into infected erythrocytes, which attain several hundred times the concentration found in plasma.

Compounds such as pentamidine, diminazene aceturate (Fig. 6.4), and ethidium bromide (Fig. 4.11), are well known trypanocides which bind to kinetoplast DNA of the flagellated trypanosome *Crithidia oncopelti*. DNA synthesis is immediately depressed while synthesis of protein and RNA continues unaffected for some time. The reason for the selective effects is unclear since the agents also bind to host DNA. The fact that blood forms of trypanosomes have no surface charge or a slight positive charge compared to the negative surface charge of host cells may allow some selective penetration into the parasite. The antimalarials mepacrine and chloroquine also bind to DNA but it is now thought that the major antiprotozoal activity of these compounds is related to their effects on membranes as mentioned above.

Metronidazole (Fig. 6.5) is important in the treatment of amoebic dysentery and trichomoniasis. More recently it has been found particularly useful to treat infections caused by anaerobic bacteria such as bacteroides, fusobacteria and clostridia, and is used to minimize sepsis after abdominal surgery. Its biochemical effects have been studied using *Trichomonas vaginalis*. The organism rapidly absorbs metronidazole and reduces it to a polar product

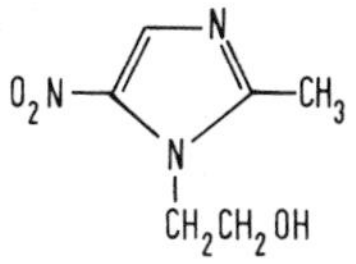

Figure 6.5 Metronidazole.

under anaerobic conditions. The as yet unidentified cytotoxic product of metronidazole (probably a nitro radical anion) is short lived but kills the cell by damaging the DNA. When oxygen tension is increased the activity of metronidazole is diminished, possibly because it competes with oxygen for the same reduced carrier in the subcellular particles called hydrogesomes. The cytotoxic product acts as a nuclease and causes DNA cleavage with the formation of thymine and thymine-derived products. Both *Trichomonas vaginalis* and *Entamoeba histolytica* have DNA with a high thymine content which renders them particularly sensitive to metronidazole because of the increased number of target sites.

COMPOUNDS USED TO TREAT FUNGAL DISEASES

Almost all organic material is subject to attack or spoilage by fungi, and hence fungicides are of great practical importance as preservatives. Many compounds are used for this purpose. The commonest types are based on heavy metals or are relatively labile sulphur compounds such as dithiocarbamates. These are generally toxic agents with multiple effects, including disturbance of membrane function and interference with respiration. Some of these compounds find use against the many fungal diseases which affect plants, although in recent years more specific compounds have been developed.

In crop protection it is a great advantage to use chemicals which are absorbed and translocated within the plant. The antifungal action can then reach parts shielded from the spray and is not lost through rain. Similarly the advent of systemically active relatively specific antifungal agents for human use has enabled the clinicians to treat the potentially dangerous deep-seated fungal infections.

The design of antifungal compounds for use in animals and plants is a demanding challenge. Fungi, unlike bacteria, are eukaryotic organisms so that many of the specific targets available in bacteria do not exist in fungal pathogens. The majority of successful antifungal agents exploit a single major difference between fungal and host cells, namely that of the composition of the cell membrane. Other agents have effects in the nucleus or against the fungal cell wall.

Sterol biosynthesis and function

POLYENES

The polyene macrolide antibiotics comprise about 200 compounds produced by *Streptomyces* spp. Of these only a few are sufficiently non-toxic to be used clinically and only one is used to treat systemic infections in man — amphotericin B (Fig. 6.6). Polyenes are active against yeasts and fungi.

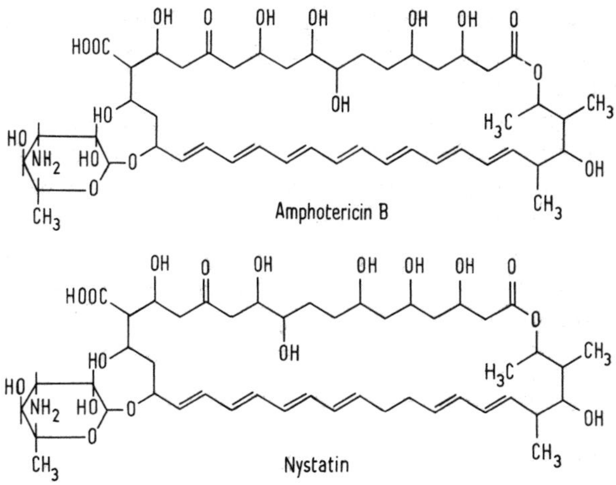

Figure 6.6 Two polyene antifungal agents.

Although infections by these organisms are rare, they are dangerous when they occur and amphotericin B can halt infections which would otherwise be fatal. The polyenes are not absorbed from the gastrointestinal tract but are sometimes given by mouth to combat fungal growth in the gut. This most commonly results from the use of broad-spectrum antibacterials which deplete the normal bacterial flora of the gut and allow yeasts and fungi to multiply and become an opportunistic infection.

The primary site of interaction is the fungal sterol, ergosterol. The sterol composition of the fungal membrane is important in determining the sensitivity to the polyene antibiotics and it is the difference in their relative affinities for ergosterol-containing fungal membranes and cholesterol-containing mammalian membranes that allows polyenes to be used clinically at all. The similarity of cholesterol and ergosterol also means that safety margins are low with polyenes. The compounds are inherently toxic to mammals, and the side effects of amphotericin B, such as kidney damage, have prevented its greater use.

Bacteria are not affected by amphotericin B but comparative studies have been carried out on the mycoplasma *Acholeplasma laidlawii*. These cells do not need sterols in their membranes, but can incorporate them, either cholesterol or ergosterol, if they are added to the growth medium. Various studies have shown that ergosterol-containing organisms are sensitive to amphotericin B whereas cells grown without sterols are not affected. Addition of ergosterol or digitonin to yeast cell cultures prevents amphotericin from being toxic; this results from the complexing of amphotericin by these agents. Mutant yeasts have been produced in the laboratory which have a block at some stage of ergosterol synthesis. These yeasts are resistant to amphotericin because there is no longer a target for the polyene in the cell. This type of resistance has not proved to be a clinical problem.

Phenotypic resistance to polyenes has been shown to develop in stationary phase and ageing yeast cultures, i.e. much more drug is required to kill cells. The concentration required for lethality has been shown to be much higher than clinically attained levels. This kind of resistance is the result of cell wall changes associated with glucans. The effect may be important *in vivo*. 'Old' cells, present in the centre of abscesses, would not be eradicated, and may be able to grow following cessation of therapy.

Polyenes bind specifically to sterols having a 3β-hydroxyl group and a relatively long side chain. The larger polyenes, nystatin (Fig. 6.6) and amphotericin B, show a preferred interaction with 3β-hydroxysterols having a conjugated diene in the nucleus. This confers some selective action against the fungi, which normally contain ergosterol, compared with erythrocytes and other body cells which have cholesterol in their membranes.

The action of compounds of this group is to increase the permeability of the fungal membrane. The effects of individual polyenes on membrane permeability show considerable differences. Filipin, for example, causes gross disruption of membranes with release of both low-molecular-weight solutes and small proteins whereas N-succinylperimycin more specifically induces the release of intracellular K^+ ions. The clinically useful drugs amphotericin B and nystatin collapse the proton motive force across the membrane by permeabilizing it to protons. The loss of K^+ from the cell may be secondary to the destruction of the proton motive force since the latter is essential for the maintenance of a high intracellular K^+ concentration relative to the extracellular environment. Living cells cannot survive a catastrophic loss of intracellular potassium so that the interaction of polyenes with cell membranes soon results in cell death.

There is little doubt that the ion-permeability enhancing effects of polyenes are caused by the drug molecules creating pores in the membranes. Molecular models of amphotericin B and nystatin show a rod-like structure held rigid by the all-*trans* extended conjugated system which is equal in length to an ergosterol molecule. The cross-section of the polyene structure is roughly rectangular. One surface of the rod is hydrophobic and the opposite surface, studded with axial hydroxyl groups, is hydrophilic. At one end of the rod the mycosamine sugar group and the carboxyl group form a zwitterionic assembly with strongly polar properties. In models it is possible to assemble about ten molecules of amphotericin B with their long axes parallel to form a cylinder. This has the hydrophilic groups on the inside and the hydrophobic groups on the outside, with one sterol molecule tucked in between each pair of molecules on the hydrophobic surface. This would constitute a pore of the correct internal diameter (0.7 nm) having a length equal to half the width of the membrane. Such a structure would accommodate itself in the membrane with the polar head group at the surface and the lipophilic outer cylinder in the lipid interior. Two such structures end to end would span the membrane and provide the observed permeability. The idea is speculative and definitive evidence for such pores in natural membranes is lacking. Furthermore, the formation of pores does not satisfactorily explain the gross permeability changes brought about by polyenes such as filipin. The nature of the interaction of polyenes with cell membranes will no doubt continue to intrigue research workers for some time to come.

ERGOSTEROL BIOSYNTHESIS

Several groups of compounds are effective antifungal agents through their actions on the biosynthesis of ergosterol. Ergosterol plays a similar role in fungal membranes to that taken by cholesterol in mammals. An outline of the biosynthetic pathway from squalene to ergosterol is shown in Fig. 6.7 and indicates the points of inhibition of several types of antifungal agents.

A group of compounds in common use includes those which inhibit at the stage of C-14 demethylation – these are generally referred to as the azole antifungals, by virtue of their nitrogen-containing heterocycle. Typical examples are the topically active agent miconazole (Fig. 6.8), which is effective against thrush and dermatophyte infections, and ketoconazole (Fig. 6.8), which is orally active and was used to treat a wide range of fungal infections particularly deep-seated mycoses. However, ketoconazole is no longer much used for trivial infections because of the danger of toxic affects on the liver and on steroid hormone biosynthesis. Azole-containing compounds have also been used in agriculture for treating various fungal infestations of plants, e.g. diclobutrazole and fenarimol. In Fig. 6.8 are examples of different chemical types with the same biochemical mode of action.

The azoles have a common mode of action involving the interaction of the lone pair of electrons on the ring nitrogen with the haem group of the cytochrome P_{450} of the enzyme catalysing the C-14 demethylation reaction. This interaction, stabilized at the binding site by the hydrophobic parts of the molecule, prevents the oxidation of the methyl group and hence its subsequent removal. The inhibition is non-competitive for the substrate and leads to a greater net reduction in flow through the metabolic pathway than competitive inhibition. The result is an accumulation of methylated sterols in the cell, with a reduction of the ergosterol content. Methylated sterols are more bulky than ergosterol and do not easily fit into a normal membrane structure. This interference in the membrane structure is thought to have adverse effects on membrane-bound enzymes, either directly on their activity or on their control. An indirect effect of ketoconazole apparently allows uncontrolled synthesis of chitin in yeasts.

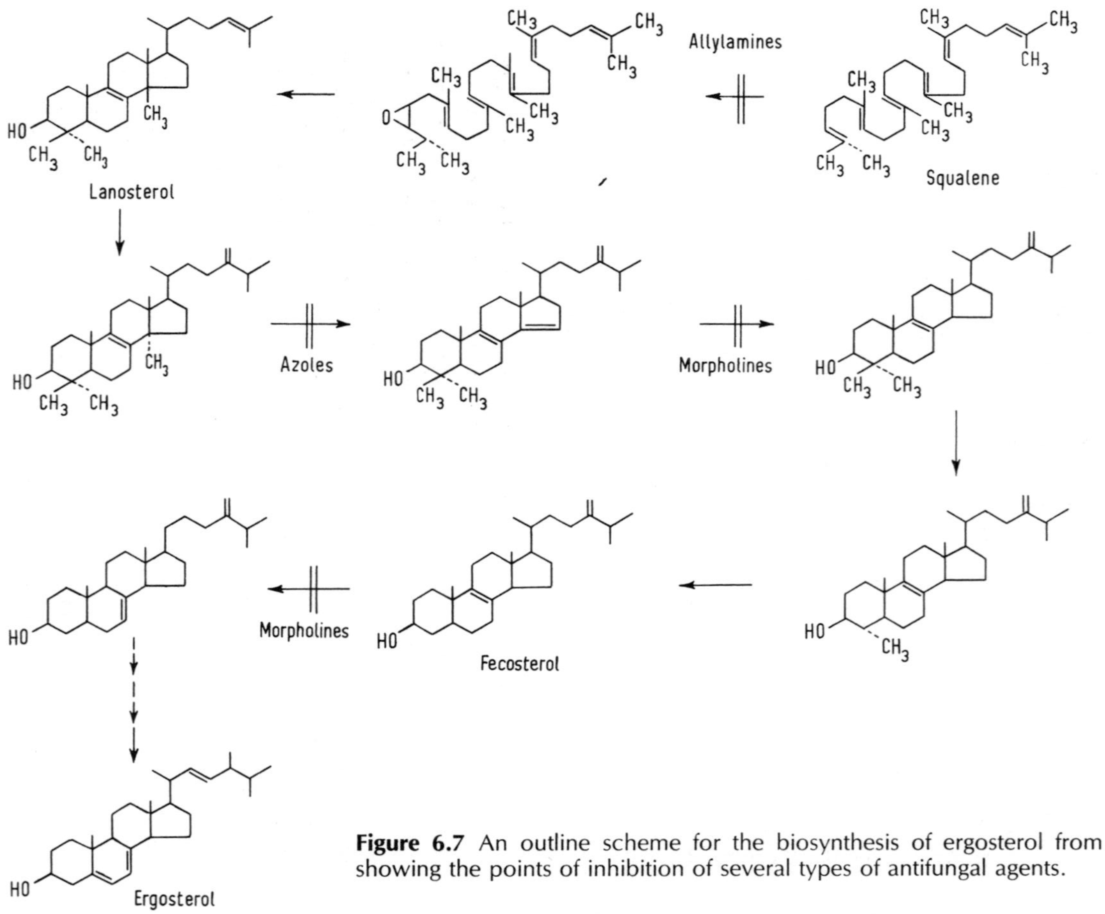

Figure 6.7 An outline scheme for the biosynthesis of ergosterol from squalene, showing the points of inhibition of several types of antifungal agents.

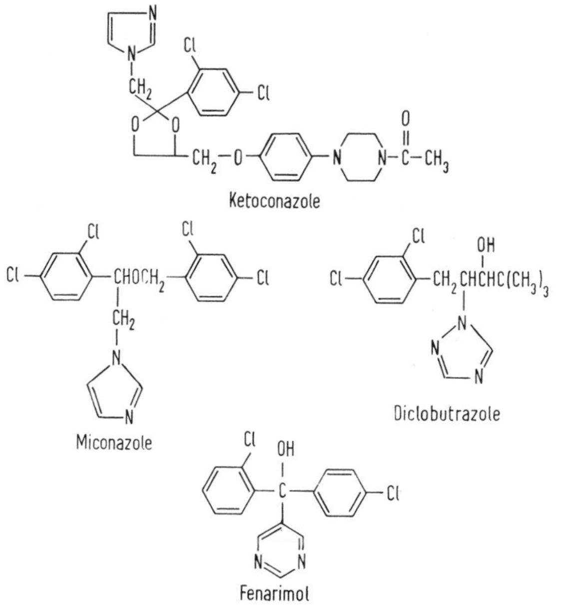

Figure 6.8 Azole antifungal agents.

A second group of compounds which are used as agricultural fungicides, but only experimentally in animals, are the morpholines, exemplified by tridemorph and fenpropimorph (Fig. 6.9). These compounds interfere with two stages of the ergosterol biosynthetic pathway. The first target is the reduction of the double bond at the 14–15 position which is formed after the removal of the C-14 methyl group. The second target is the isomerization of the double bond between carbon atoms 8 and 9 of fecosterol to a position between carbons 8 and 7. The balance between these two inhibitory activities

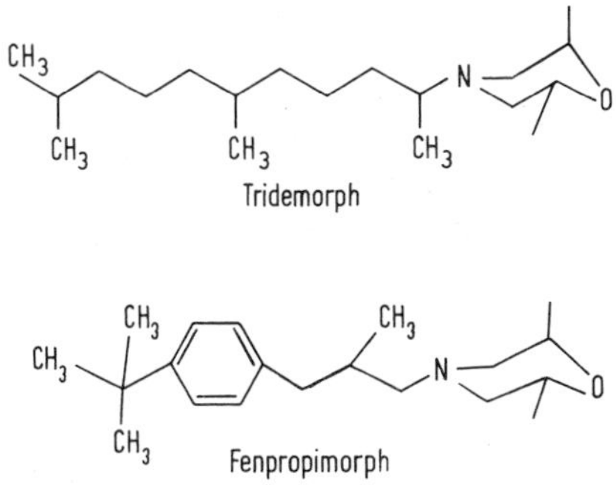

Figure 6.9 Morpholine derivatives that are antifungal agents.

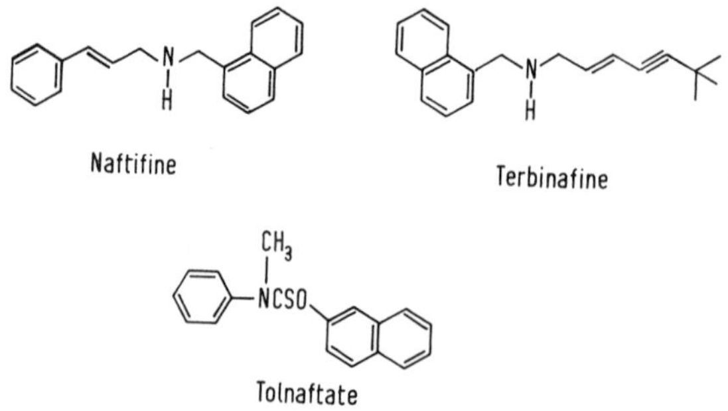

Naftifine

Terbinafine

Tolnaftate

Figure 6.10 Antifungal agents that inhibit squalene expoxidation.

varies from fungus to fungus and probably reflects subtle differences in the enzymes involved. Tridemorph inhibits *Ustilago maydis* mainly at the C-14 reduction step whereas *Botrytis cinerea* is inhibited mainly at the C-8–C-7 isomerization. The end result of either inhibition is probably the same – incorrectly formed membranes, because of inadequate supplies of ergosterol.

A third group of compounds used clinically in the treatment of dermatophyte infections affects ergosterol biosynthesis at an earlier stage. The allylamine antifungal agents, naftifine (Fig. 6.10), a topical agent, and terbinafine (Fig. 6.10), an orally active agent, inhibit the epoxidation of squalene. Recently tolnaftate (Fig. 6.10) a topical antidermatophyte compound which has been in use for many years has been shown to have the same mode of action. Squalene accumulates in the cell with a reduction in cellular ergosterol content and growth inhibition could result from either action. These compounds are much less active against yeasts although in broken cell preparations of *Candida albicans* squalene epoxidation can be inhibited. The difference in efficacy may be simply that yeast enzymes are less susceptible or perhaps yeasts have a lower sensitivity to the consequent build-up of squalene.

Antifungals which inhibit ergosterol biosynthesis show marked selectivity for fungal systems. The azole antifungals are several hundred times more potent against lanosterol demethylation in fungi than the corresponding reaction in mammals. Similarly, naftifine is several hundred times more potent against fungal squalene epoxidase. This species selectivity is critical in making the ergosterol biosynthesis inhibitors such good antifungal agents.

Nuclear function

Griseofulvin (Fig. 6.11) is a naturally occurring antibiotic. It is given orally to treat ringworm and other fungal infections of the skin and nails. In fungi

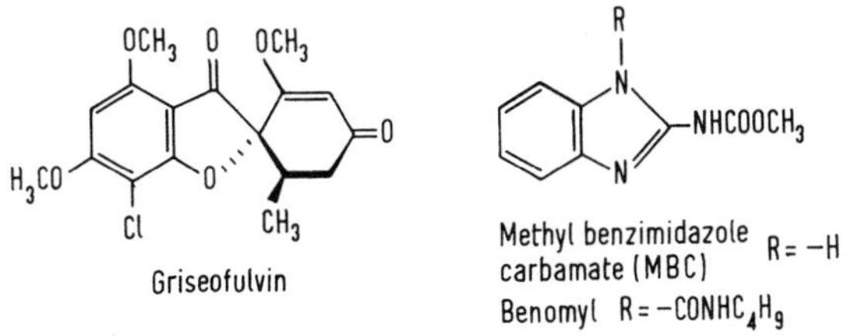

Figure 6.11 Agents that affect nuclear function in fungi.

sensitive to griseofulvin the drug causes the tips of the hyphae to become curled and growth ceases. Further study has shown that griseofulvin binds to proteins involved in the assembly of microtubules at a stage which is vital to the sliding of microtubules during the separation of chromosomes in cell division. The selectivity of griseofulvin is not fully understood since microtubules are common to all cells. However there are over 100 differences between the amino acid sequences of the tubulin of *Saccharomyces pombe* and pig tubulin. The amino acid sequences of tubulins within the animal kingdom are well conserved. The antifungal specificity of griseofulvin may well depend therefore on a differential ability to interact with the tubulins of fungi and mammalian tubulins.

Benomyl (Fig. 6.11) acts against a wide range of phytopathogenic fungi and is used extensively in agriculture and horticulture. It is metabolized in plants to methyl benzimidazole carbamate (MBC) (Fig. 6.11) which is believed to be the active agent. MBC also affects the separation of chromosomes in mitosis; it binds directly to the tubulin monomers and interferes with microtubule assembly but at an earlier stage than griseofulvin.

5-Fluorocytosine (5FC) (Fig. 6.12) was originally synthesized as an anti-cancer agent but it is now used mainly against a few medically important yeasts. 5FC, like other drugs based on nucleotide precursors, is not active *per se* but is metabolized to another compound which is the active drug. It is first deaminated to 5-fluorouracil, converted to the nucleoside triphosphate, and then incorporated into fungal RNA. In addition, 5-fluorouracil is converted to 5-fluoro-2-deoxyuracil which inhibits thymidylate synthase and therefore DNA biosynthesis. This allows a second antifungal activity. The efficacy of

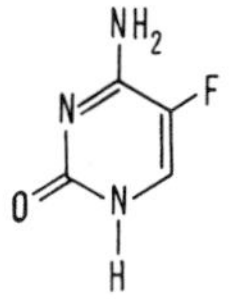

Figure 6.12 5-Fluorocytosine.

the 5FC results from a combination of inhibition of nucleic acid synthesis and miscoding caused by the incorporation of 5-fluorouracil into RNA.

Resistance to 5FC arises commonly in clinical use because several enzymes are necessary for its action, each of which may mutate. These include cytosine permease, cytosine deaminase and UMP pyrophosphorylase. During therapy with 5FC careful monitoring of blood levels is required to ensure that concentrations toxic to the kidney are not achieved and also to ensure that levels do not fall to concentrations which allow resistance to develop.

Inhibition of the biosynthesis of the fungal cell wall

The outer wall of the fungal cell is a complex multilayered structure where amorphous, granular and fibrillar structures interact with one another to give the cell a rigid shape and to confer osmotic stability. As with bacteria, the cell wall is therefore a major target for chemotherapy. The cell walls of fungi do not contain peptidoglycan so that the β-lactam antibiotics have no effect. In recent years research to discover agents that affect the fungal cell wall has been intense but although several compounds have emerged none has so far demonstrated clinical effectiveness.

Certain natural products of streptomycetes, exemplified by polyoxin D and nikkomycin Z (Fig. 6.13), are potent inhibitors of chitin synthesis. In fungal walls the rigid structural framework is provided by this polymer of N-acetylglucosamine. The polyoxins consist of a pyrimidine nucleoside and a peptide moiety. Polyoxin D is a potent competitive inhibitor of chitin synthase, a membrane-located enzyme which catalyses the transfer of N-acetylglucosamine units from UDP-N-acetylglucosamine to a chitin chain. The competition is clearly due to the structural analogy between the drug and the natural substrate. The K_i value for chitin synthase from *Mucor rouxii* is 0.6 βM whilst the K_m for UDP-N-acetylglucosamine is 0.5 mM. Inhibition of chitin synthesis in yeasts blocks septum formation whilst irregular hyphae are formed in *Candida albicans*. Treatment of the plant pathogen *Altenaria kikuchiana* with polyoxin inhibits the incorporation of glucosamine into chitin resulting in abnormal accumulation of UDP-N-acetylglucosamine. The growing hyphal tips swell and undergo osmotic lysis. Filamentous fungi are more susceptible than yeasts probably because of their greater chitin content and their reliance on its strength for osmotic stability.

The polyoxins rely for their antifungal activity on transport into the cell by a permease system that normally carries dipeptides into cells. The loss of this peptide permease renders the cell resistant to polyoxins. This happened in Japan where polyoxin D was used to control blackspot on peaches caused by *Altenaria* spp. The loss of the peptide permease does not appear to disadvantage the cells. The presence of peptides in body fluids, which compete with polyoxins for the fungal permease, excludes any useful antifungal activity in animals and man.

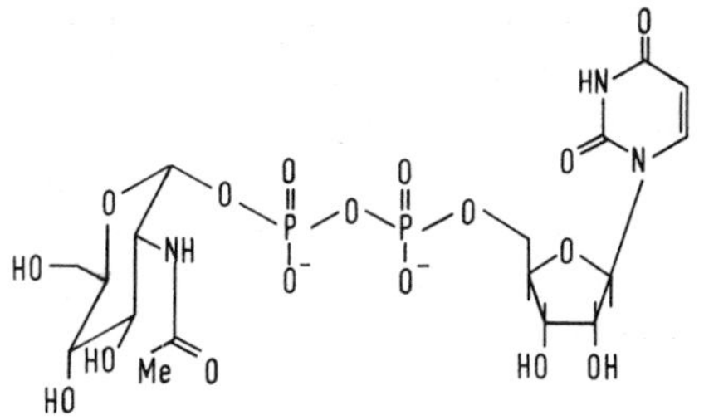

UDP-*N*-acetyl-D-glucosamine

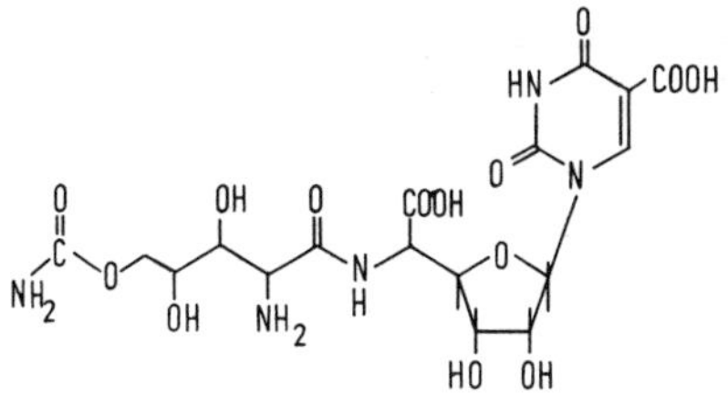

Polyoxin D

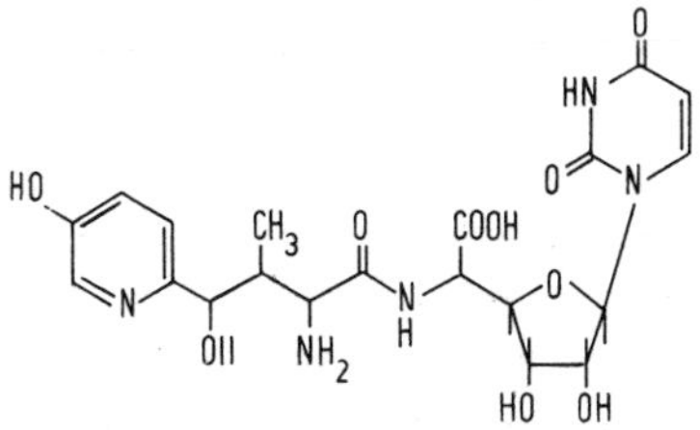

Nikkomycin Z

Figure 6.13 Antifungal agents that inhibit cell wall chitin synthesis, together with the substrate UDP *N*-acetylglucosamine.

Several cyclic peptides containing a fatty acid chain which also interfere with cell wall biosynthesis are very active antifungal compounds, mostly against pathogenic yeasts. Echinocandin B (Fig. 6.14) belongs to this category. Echinocandin B causes lysis of the budding yeast cells at the growing tip and of the elongating hyphae of *Candida albicans*. This compound causes thinning and disorganization of the constituent layers of the cell wall, suggesting a blockade of some step in cell wall construction. Echinocandin inhibits biosynthesis of the $\beta 1{:}3$ glucan wall polymer that confers overall rigidity and plays a role equivalent to chitin in filamentous fungi.

Energy metabolism

Several antimicrobial agents inhibit respiration in fungi. With some compounds the effect is secondary to other actions such as disturbance of membrane function, but with others the action is a primary effect on the enzyme systems that transfer hydrogen from the dehydrogenases to molecu-

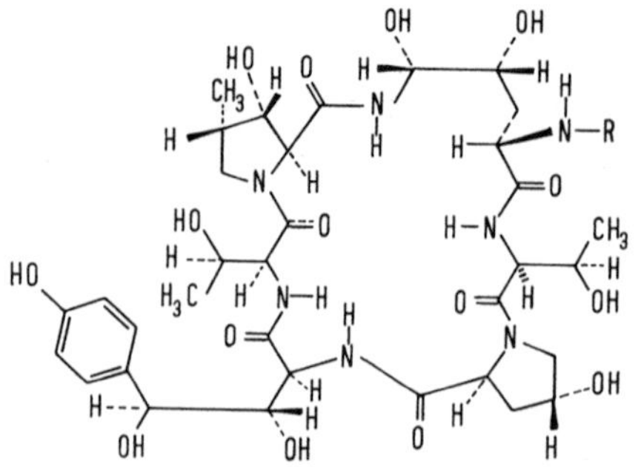

Figure 6.14 Echinocandin B: inhibitor of the biosynthesis of the glucan cell wall polymer in yeasts.

lar oxygen. The compounds that possess this type of action generally lack selectivity towards micro-organisms; their effects on animal cells mean that they are toxic and of little use as therapeutic agents. However, they have proved to be extremely useful biochemical tools and will therefore be considered in more detail.

Antimycin (Fig. 6.15) inhibits the growth of yeasts and many fungi but has little effect on bacteria. Cells that are affected by antimycin show an immediate fall in oxygen uptake. Subcellular particles from the rice blast fungus *Piricularia oryzae* containing the cytochrome system also show inhibition of respiratory function. Low concentrations of antimycin inhibit oxygen uptake by the yeast *Saccharomyces cerevisiae* and stimulate fermentation. The organism thus responds as though it were deprived of oxygen. All these results point to an action on the cytochrome pathway and this supposition is fully confirmed by other work on subcellular preparations from animal cells.

The point of action of antimycin has been mainly studied using well-defined submitochondrial particles from mammalian cells, but it seems likely that fungal mitochondria are similarly affected. The sequence of reactions between the dehydrogenases and molecular oxygen is set out in Fig. 6.16. Antimycin inhibits the respiration of unfractionated particulate oxidase preparations but has no effect on isolated dehydrogenases. The point of attack is further defined by the finding that in these same preparations antimycin inhibits the reduction of cytochrome c by NADH but does not inhibit the oxidation of reduced cytochrome c by molecular oxygen. The differential spectroscopic method enables the state of oxidation or reduction of all the cytochrome components to be ascertained in a respiratory enzyme preparation. The block comes immediately before cytochrome c_1. This is further confirmed by the demonstration that antimycin inhibits the isolated system ubiquinolcytochrome c oxidoreductase (complex III). The reason for

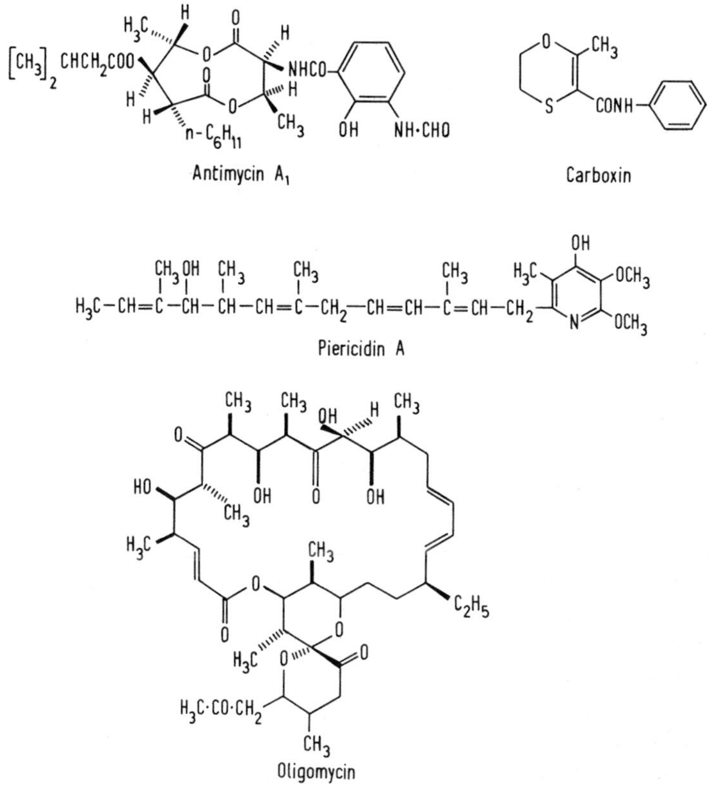

Antimycin A$_1$

Carboxin

Piericidin A

Oligomycin

Figure 6.15 Antifungal agents that interfere with energy metabolism.

the block is not known in molecular terms though an allosteric type of effect on the mitochondrial enzyme complex has been suggested.

Carboxin (Fig 6.15) is a synthetic fungicide used as a seed dressing to control smut infections in cereal crops. It acts by blocking the succinate branch of the respiratory pathway (Fig. 6.16) at the iron–sulphur protein of the dehydrogenase complex. Piericidin (Fig. 6.15) is a fungicidal and insecticidal antibiotic which specifically blocks a similar region in the other branch of the respiratory chain serving the NAD-dependent dehydrogenases. Piericidin has no commercial application but, like antimycin, is a valuable biochemical tool for studies on the respiratory pathway.

Oligomycin (Fig. 6.15) and the closely related rutamycin strongly inhibit a limited number of yeast and fungal species (hence oligomycin). They also act on cells of animal origin, but not on bacteria.

Oligomycin reduces the oxygen consumption of intact cells and also of some mitochondrial preparations. However, unlike antimycin its effect is

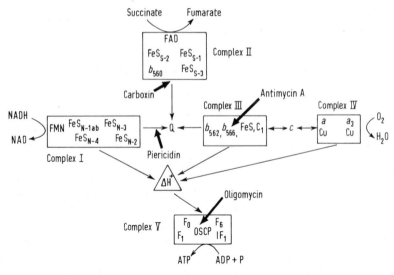

Figure 6.16 Profile of the components of the respiratory chain, showing points of inhibition of antifungal agents. a, b, c are cytochromes; FeS are iron–sulphur clusters; Q is ubiquinone; ΔH^+ represents the energy potential; OSCP (the oligomycin sensitivity-conferring protein), F_0, F_1, F_6 and IF_1 are component proteins of Complex V.

only evident in strongly coupled mitochondria supplied with inorganic phosphate and ADP. If 2,4-dinitrophenol or some other uncoupling agent is added to such preparations, respiration continues even in the presence of oligomycin. The antibiotic has no effect on the oxygen consumption of submitochondrial particles which lack the oxidative phosphorylation mechanism. These observations indicate that the effect of oligomycin on respiration is due to inhibition of the oxidative phosphorylation stages which are normally coupled to respiration in intact cells.

Oxidative phosphorylation is associated with a particular adenosine triphosphatase bound to the mitochondrial membrane. Particulate preparations of this enzyme, Complex V, are strongly inhibited by concentrations of oligomycin ($1\,\mu g\ ml^{-1}$) which just inhibit fungal growth. They not only catalyse the breakdown of ATP but also the reverse reaction:

$$ADP + P_i + H^+ \rightleftharpoons ATP + H_2O$$

and ATP–P_i exchange.

Oxidative phosphorylation involves membrane function and the effect of oligomycin depends on its ability to decrease the proton permeability of the membrane. Such a decrease, by 5–10 times, has been shown experimentally in mitochondrial particles from beef heart and in chromophores from *Rhodospirillum* treated with oligomycin.

COMPOUNDS USED TO TREAT VIRUS DISEASES

The success of drugs in the treatment of bacterial infections contrasts strongly with the relatively meagre returns from the search for compounds active against virus infections. This is not surprising in view of the difference in the way these agents multiply in the body. Bacteria grow independently and have their own peculiar metabolic characteristics which provide points of selective attack. The process of virus replication, on the other hand, is essentially a subversion of normal cell metabolism and the changes involved are subtle. It is therefore difficult to find compounds which will prevent virus replication without damaging some vital cellular mechanism. Nevertheless, virus infection and reproduction are complex processes which present possibilities for interference at various stages. These can be considered in three groups. Firstly the stage of viral infection which involves the initial binding of virus to the target cell, followed by penetration into the cell and virus uncoating. The second phase concerns the biosynthesis of viral macromolecules. Finally, there is the maturation of virus particles and their packaging. Despite considerable efforts, very few compounds are currently accepted as having a useful effect against diseases caused by viruses.

Initial phase of cell–virus interaction

Amantadine (Fig. 6.17) has both prophylactic and curative effect against influenza A_2 infections in mice. In man it certainly offers protection against influenza A_2 infection during the period of dosing, but curative action on the established disease is doubtful. There is evidence that amantadine acts by preventing the penetration of the virus into the host cell; it may also inhibit the uncoating of the virus particle. Washing cells which have been exposed to amantadine to remove the drug eliminates its protective effect even though

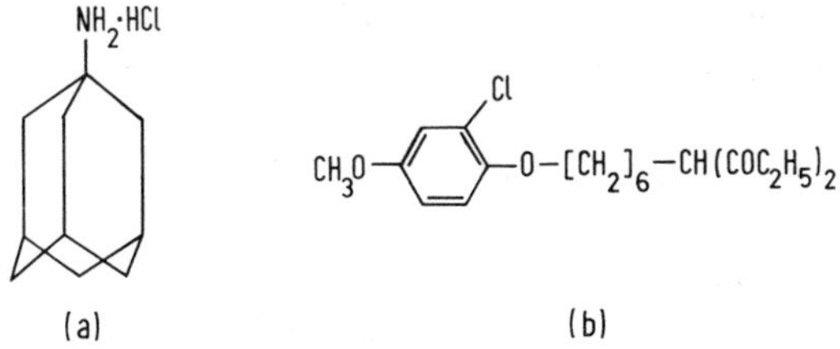

(a)   (b)

Figure 6.17 (a) Amantadine, an anti-influenza agent preventing virus access to the cell. (b) Arildone, an antiviral compound that blocks the uncoating of virus particles.

amantadine remains inside the cells. This result indicates that the drug's main site of action is probably at the cell membrane. Human immunodeficiency virus (HIV), which causes AIDS, contains large amounts of lipid with a cholesterol/phospholipid ratio of unity. A mixture of lipids known as AL721 when added to cell cultures infected with HIV reduces the cholesterol/phospholipid ratio of the virus. This induces changes in membrane microviscosity and the conformation of viral membrane proteins. Viral infectivity is reduced probably as a result of diminished virus uptake.

Arildone (Fig. 6.17) prevents the uncoating of several RNA and DNA viruses by interacting intracellularly with the capsid proteins. This results in inhibition of virus replication at concentrations which have no effect on host cells. Arildone treatment of mice infected with polio virus prevents their death, and the development of lesions in guinea pigs infected with herpes simplex virus can be prevented by local treatment with the compound.

Macromolecular biosynthesis

In most viruses the genome is replicated by the virus-specific polymerases and in retroviruses by reverse transcriptase. The genome is transcribed into messenger RNA from which viral proteins are translated. In addition to the polymerases other virus enzymes are needed for post-transcriptional modification of mRNA. These enzymes participate in polyadenylation, methylation and capping, each of which is a potential target for antiviral action.

ACYCLOVIR

To date the most successful clinically active antiviral agent that interferes with macromolecular synthesis is acyclovir (Fig. 6.18). This is a purine nucleoside analogue lacking the ribose ring structure. It is used extensively for many conditions caused by herpes (DNA) viruses and can be given topically, orally or by injection. Acyclovir is in reality a 'pro-drug'; it must first be activated to the monophosphate by thymidine kinase and then to the triphosphate by thymidylate kinase. The triphosphate of acyclovir inhibits viral DNA polymerase 100 times more effectively than the host polymerase. Acyclovir triphosphate is also a substrate for the viral polymerase and causes DNA chain termination since it lacks the 3'-OH group of deoxyribose. Acyclovir triphosphate is in contrast, a poor substrate for the host DNA polymerase.

OTHER ANTIVIRAL NUCLEOSIDE ANALOGUES

Several of these compounds exert clinically useful effects and all are converted successively to the mono-, di- and triphosphates by intracellular

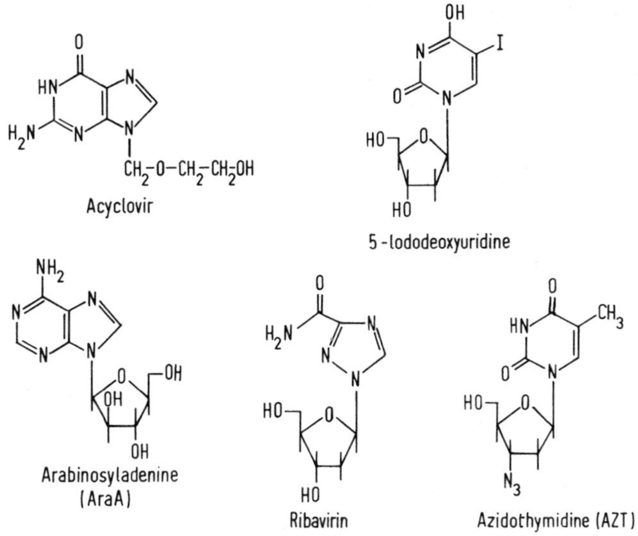

Figure 6.18 Antiviral nucleoside analogues.

enzymes. In 5-iododeoxyuridine (Fig. 6.18) the similar van der Waals radii of the iodine atom (0.215 nm) and the methyl group (0.2 nm) of thymidine enable the drug to replace thymidine in DNA with considerable efficiency since its triphosphate is accepted as a substrate by DNA polymerase. The incorporation of 5-iododeoxyuridine into viral DNA leads to errors of replication and transcription and the eventual termination of virus replication. The compound has been used topically to treat herpes simplex infection of the cornea.

Vidarabine, 9-β-D-arabinosyladenine (AraA) (Fig. 6.18) inhibits several viruses including herpes, cytomegalovirus and the Epstein-Barr virus by its 5′-triphosphorylated metabolite. This leads to inhibition of viral DNA polymerase, viral ribonucleotide reductase and to direct incorporation into viral DNA. All of these effects occur at concentrations of AraA below those required to inhibit host DNA synthesis.

Ribavirin, 1-β-D-ribofuranosyl-1,2,4-triazole-3-carboxamide (Fig. 6.18) is active *in vitro* against a wide range of RNA and DNA viruses. Ribavirin 5′-triphosphate blocks the capping of virus-specific RNA by the inhibition of messenger RNA guanylyl transferase activity. This mechanism may be important in the inhibition of the influenza RNA virus by ribavirin.

Azidothymidine, 3′-azido-3′-deoxythymidine (Fig. 6.18) or AZT as it is commonly known, is the most successful drug developed so far for the treatment of AIDS. It strongly inhibits the replication of the HIV retrovirus in T lymphocytes. The phosphorylated metabolites of AZT exert several points of inhibition in viral replication: the monophosphate inhibits thymidy-late kinase thus restricting the availability of thymidine triphosphate for viral replication. The triphosphate of AZT is both an inhibitor and a substrate for HIV-specific reverse transcriptase and when incorporated into the viral DNA serves as a chain-terminating agent. AZT is not entirely specific for HIV replication and its ability to induce pyrimidine starvation probably contributes to the serious side effect of bone marrow suppression.

Maturation of viruses

Most emphasis on virus chemotherapy has inevitably been directed against nucleic acid synthesis, but the maturation and packaging of viruses probably provide alternative targets. These processes require the post-translational modification of capsid proteins by specific proteases or glycosylation systems. A modified amino acid carbobenzoxyleucylchloromethylketone (ZLCK, Fig. 6.19) is an inhibitor of virus proteases and in laboratory experiments inhibited the cleavage of polio virus precursors and reduced the yield of virus.

Methisazone (Fig. 6.19) prevents the development of encephalitis in mice infected with various pox viruses. It was claimed to have prophylactic action in human smallpox, but now this disease has been eliminated by mass vaccination programmes it is no longer used. Nevertheless its antiviral action is of interest. Studies in cultured cells infected with vaccinia virus show that the compound inhibits the synthesis of the protein required for encapsidation of the mature virus, by causing a fragmentation of the viral mRNA produced

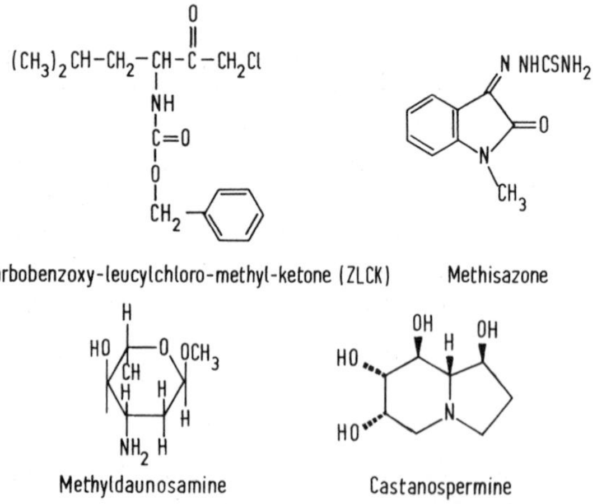

Carbobenzoxy-leucylchloro-methyl-ketone (ZLCK)   Methisazone

Methyldaunosamine   Castanospermine

Figure 6.19 Antiviral compounds which inhibit viral maturation.

late in virus development. There is no interference with the normal protein synthesis in the cell.

Several sugar analogues have been shown to have effects on virus glycoprotein biosynthesis, with the result that incorrectly formed glycoproteins are deposited on the cell surface which do not allow virus coating. The best example is methyldaunosamine (Fig. 6.19). Recently castanospermine (Fig. 6.19), a natural product from the Australian chestnut tree which inhibits β-glucosidase and other reactions involving sugars, has been shown to inhibit glycosylation of the surface glycoproteins of HIV and so prevent its replication. Analogues are currently sought with less host toxicity.

Interferon

For many years the natural lymphokine interferon was held to be the main hope for antiviral therapy. With the advent of genetic engineering large amounts of the various forms of this protein became available to test the hypothesis. Interferon is released when viruses interact with target cells. Newly synthesized interferon is secreted from the cell and binds to receptors on neighbouring cells. This interaction triggers the synthesis of at least three cellular proteins that act to prevent viral replication but have no effect on normal macromolecular synthesis. One protein is a protein kinase which activates a eukaryotic initiation factor eIF_2 needed for the synthesis of viral protein. A second protein catalyses the formation of the oligonucleotide pppA2'p5'Ap5'A which in turn activates the third protein, an endo RNase which degrades viral mRNA and prevents viral replication. Clinical studies with interferon have shown antiviral activity in man although its usefulness is limited by significant side effects. In addition interferon has some utility in enhancing the antiviral effects of agents such as amantadine, ribavirin and acyclovir.

Although the antiviral drugs now available have limited use in clinical medicine, they serve to demonstrate that viral chemotherapy is possible. The newer agents in development support this notion. Even where an active compound has been found however its action often appears to be restricted to a narrow class of viruses, or even a single strain. The advent of AIDS and HIV has led to a major increase in the search for antiviral agents, particularly against RNA viruses and viruses where viral DNA becomes incorporated into the host genome with long latency periods before the clinical signs of disease appear.

FURTHER READING

de Clerq, E. (1985) Antiviral agents. In *The Scientific Basis of Antimicrobial Therapy, Society for General Microbiology Symposium 38* (eds D. Greenwood and F. O'Grady).

Esteban, M. and Paez, E. (1985) Antiviral and antiproliferative properties of interferons: mechanism of action. *Progr. Med. Virol.*, **32**, 159.

Fromtling, R. D. (ed.) (1987) *Recent Trends in the Discovery, Development and Evaluation of Antifungal Agents*, J. R. Prous. Science.

Kerridge, D. (1986) Mode of action of clinically important antifungal drugs. *Adv. Microb. Physiol.*, **27**, 1.

Peters, W. (1985) Antiprotozoal agents. In *The Scientific Basis of Antimicrobial Therapy, Society for General Microbiology Symposium 38*, (eds D. Greenwood and F. O'Grady).

Ryley, J. F., Wilson, R. G., Gravestock, M. B. and Poyser, J. P. (1981) Experimental approaches to antifungal chemotherapy. *Adv. Pharmacol. Chemother.*, **18**, 49.

Streissle, G., Paessens, A. and Oediger, H. (1985) New antiviral compounds. *Adv. Virus Res.*, **30**, 83.

7 Penetrating the defences: how antimicrobial agents reach their targets

In order for a drug to inhibit microbial growth it has to reach an inhibitory concentration at its target site. The drug must therefore penetrate the various permeability barriers that separate the target site from the external environment. In the case of bacteria the differences in the properties of these permeability barriers among the various genera and species are important in determining the antibacterial spectrum of a drug. Target sites, e.g. ribosomes, DNA, enzymes, etc., isolated from different groups of bacteria, often exhibit similar sensitivity to an antibacterial agent whereas the intact cells may have very different sensitivities. In many cases this is due to differences in the structure and composition of the cell envelopes which control the penetration of drugs to the target sites. Resistance to some antimicrobial agents depends upon diminished cellular permeability. The change in permeability may be exquisitely selective, affecting only the movement of the inhibitor into the cell, whilst in other cases a generalized non-specific decrease in cellular permeability keeps inhibitors out. The relationship of acquired drug resistance to drug penetration is described in greater detail in Chapter 8.

CELLULAR PERMEABILITY BARRIERS TO DRUG PENETRATION

The cytoplasmic membrane

Common to all cellular micro-organisms is the permeability barrier provided by the cytoplasmic membrane. Its exact composition depends very much on the cell of origin but the most important feature is the lipid bilayer found in all cytoplasmic membranes. Drugs may cross this barrier either by passive diffusion or by facilitated diffusion involving a biological carrier system.

161

PASSIVE DIFFUSION

The rates of passive diffusion of uncharged organic molecules across lipid membranes correlate reasonably well with their lipid/water partition coefficients. That is, the more lipid-soluble a compound is, the more readily it enters and diffuses across the membrane. However, in a chemically related series of drugs, if the lipid solubility increases to a point where a drug is virtually insoluble in water, biological activity may decline sharply, presumably because the drug is reluctant to leave the lipid phase of the membrane for the watery environment of the cell interior. This phenomenon has been given mathematical expression in the 'Hansch equation' (so named after the scientist who formulated it):

$$\log (1/C) = -k(\log P)^2 + k'\log P + \varrho\sigma + k''$$

where C is the molar concentration of the drug necessary to cause a standard biological response, in this case inhibition of cell growth and division, P is the partition coefficient, ϱ and σ are physicochemical constants (Hammett constants) defining certain electronic features of the molecule and k, k' and k'' are empirically determined constants. The Hansch equation defines the parabolic relationship between the biological activity and the lipophilic character of a drug and in addition takes account of the electronic and structural features of the molecule. It indicates that in a chemically related series of biologically active drugs there is an optimal partition coefficient associated with maximum biological activity. It should be noted, however, that this relationship holds only for those agents which cross membranes by passive diffusion and it may be irrelevant when biologically facilitated transport of a drug is involved.

The Hansch equation has been applied to various sets of synthetic antibacterial compounds which penetrate the bacterial envelope by passive diffusion. The results show that compounds that are most active against Gram-negative bacteria are less lipophilic than compounds highly active against Gram-positive organisms. The cytoplasmic membranes of the two classes of bacteria are sufficiently similar in total lipid content to make it unlikely that the differences in the partition coefficient of optimally active compounds could be explained by differences in the solvent properties of the cytoplasmic membranes. A more likely explanation is to be found in the different nature of the outer envelopes of Gram-negative and Gram-positive bacteria (Chapter 2). We shall return to this important topic later in the chapter.

The rates of penetration of water-soluble molecules across a lipid membrane by passive diffusion are usually very low. Un-ionized molecules of molecular weight less than 100 move across as though the membrane were interspersed with water-filled channels or pores. These hypothetical water-filled channels of the cytoplasmic membrane must be clearly distinguished from the larger well-characterized hydrophilic pores in the outer membrane of Gram-negative bacteria (Chapter 2). While water-soluble antibiotics of molecular weight up to 600 or so pass through the pores of the outer membrane, their molecular size precludes them from using the water-filled channels of the cytoplasmic membrane. Nevertheless, as we shall see, some water-soluble antibiotics readily enter the bacterial cytoplasm.

Ionized molecules of molecular weight greater than 100 are also virtually unable to cross cytoplasmic membranes by passive diffusion. This is because ionized molecules in aqueous solution possess a strongly bound hydration shell, and only the very smallest ions can diffuse across membranes via the watery channels. The effect of ionization on the activity of an antibacterial agent is well illustrated by erythromycin. The pK_a of the basic group of this antibiotic is 8.8 and the concentration required for antibacterial activity decreases markedly as the pH of the bacterial medium is increased towards 8.8. Very likely only the un-ionized form of erythromycin penetrates into the bacteria and this represents an increasing proportion of the total erythromycin as the pK_a of the drug is approached.

FACILITATED DIFFUSION

A remarkable feature of cytoplasmic membranes is their ability to transfer certain physiologically important molecules across the membranes at rates very much higher than could occur by passive diffusion. This process, which is especially noticeable with water-soluble and ionized molecules, is known as facilitated transfer or facilitated diffusion. Characteristically, the rate of transfer of the permeant is proportional to its concentration over a limited range beyond which a limiting rate is approached. This is generally attributed to the involvement in the transfer process of carrier molecules specific for each permeant. The rate of transfer increases with increasing permeant concentration until all of the carrier sites are saturated. Facilitated diffusion can therefore usually be distinguished from passive diffusion, because in the latter process the transfer rate is proportional to the permeant concentration over a much wider range. Facilitated transfer by itself results in the equilibration of the permeant across the membrane. In some instances, however, when the transfer system is linked to an input of 'energy' usually the hydrolysis of ATP or the proton motive force across the cytoplasmic membrane, the permeant is transferred across the membrane against its concentration gradient. This is known as 'active transport'. Facilitated transfer systems are often highly specific and only close structural analogues of the natural permeant compete effectively for the transport sites. These transport systems are vitally important for the acquisition of nutrients from the extracellular environment. From this we would expect facilitated transfer and active transport to be used only by those drugs resembling natural cellular nutrients. This is generally borne out by experience, but there are examples (see below) of drugs that do not obviously resemble known nutrients which apparently gain access to their target sites by facilitated transfer.

The outer layers of bacterial cells

Although the cytoplasmic membrane is a most important barrier against the penetration of many water-soluble antimicrobial chemicals, it is unlikely to provide the basis for the marked specificity of action of some antibiotics. Table 7.1 lists a number of agents that are active against both Gram-positive and Gram-negative bacteria and also several that are significantly active only against Gram-positive organisms. The poor activity of this latter group of drugs against Gram-negative bacteria is almost certainly due to poor penetration of the complex outer layers of Gram-negative cells.

The structures of the cell envelopes of the two main bacterial groups are described in Chapter 2. In Gram-negative bacteria the thin peptidoglycan layer is probably not a serious obstacle to the entry of compounds, but the outer membrane presents a formidable permeability barrier to many antibacterial agents. Because of their strongly polar, predominantly negatively charged nature, the teichoic acids of Gram-positive cells may influence the penetration of ionized molecules. The interaction of positively charged compounds, such as streptomycin, with teichoic acid generates a localized

Table 7.1   Differential sensitivity and permeability to typical antibacterial drugs

Drugs active against Gram-positives and Gram-negatives	Drugs less active against Gram-negatives because of permeability barriers
Tetracyclines	Benzylpenicillin (penicillin G)
Streptomycin and aminoglycosides	Methicillin
Albomycin	Macrolides
Sulphonamides	Lincomycin
D-Cycloserine	Rifamycins
Chloramphenicol	Fusidic acid
Fosfomycin	Vancomycin
Puromycin	Bacitracin
Many synthetic antiseptics	Novobiocin
Nitrofurans	Actinomycin D
Ampicillin and carbenicillin	
Thienamycin	

high concentration of drug in the envelope which may enable the drug to challenge the permeability barrier of the cytoplasmic membrane more effectively. In contrast, the entry of anionic molecules may be retarded by teichoic acid, although the exquisite sensitivity of many Gram-positive bacteria to penicillins which are organic anions suggest that the repulsive effect of teichoic acid cannot be very significant.

The first indication that the layers lying outside the cell membrane influence drug penetration came from studies with Gram-negative cells with defective envelopes. L-Phase (or L-Forms) of *Proteus mirabilis* were found to be 100 to 1000 times more sensitive than intact cells to erythromycin and several other macrolides. There was a smaller increase in sensitivity to other antibiotics including streptomycin, chloramphenicol and the tetracyclines. In L-forms the peptidoglycan envelope is also defective so that the relative contributions of the various outer layers of intact bacteria to the barrier function remained uncertain. Many other studies since have clearly demonstrated the importance of the outer membrane as the major permeability barrier.

Treatment of Gram-negative bacteria with the chelating agent ethylene-diaminetetraacetic acid (EDTA) releases 30–50% of the surface lipopolysaccharide, probably by sequestering the divalent cations, Ca^{2+} and Mg^{2+}, which stabilize the outer envelope. Gram-negative cells treated in this way become much more sensitive to a number of agents, including lysozyme and the antibiotics listed in Table 7.1 to which intact Gram-negative cells are intrinsically resistant. Increased uptake of several antibiotics into EDTA-treated cells has been noted and since the cytoplasmic membrane and peptidoglycan are hardly affected by this treatment, changes in the outer membrane and a loss of the associated lipopolysaccharide are believed to be largely responsible for the increased permeability and antibiotic sensitivity.

The lipopolysaccharide is confined mainly to the outer leaflet of the outer membrane and confers a hydrophilic character on this structure which hydrophobic drugs find difficult to breach. It is not surprising therefore that loss of lipopolysaccharide caused by physical means or by mutations, as in the so-called 'deep rough' mutants of *Salmonella* and *Escherichia coli*, increases bacterial sensitivity to hydrophobic drugs. Mutants primarily selected for higher sensitivity to these drugs usually have alterations in the chemical constitution of the lipopolysaccharide in the outer membrane that presumably adversely affect its barrier function. *Neisseria gonorrheae* is unusual among Gram-negative bacteria in being intrinsically sensitive to hydrophobic antibiotics, perhaps because the outer membrane lipopolysaccharide provides an inadequate hydrophilic envelope for the organism. In general, however, hydrophilic antibiotics gain access to the outer layers of Gram-negative bacteria more readily than hydrophobic agents.

PORE STRUCTURE OF THE OUTER MEMBRANE OF GRAM-NEGATIVE BACTERIA

Penetration of the outer membrane of Gram-negative bacteria by hydrophilic molecules is not only facilitated by the lipopolysaccharides but also by a remarkable group of proteins called *porins*. The porins, described in Chapter 2, are arranged to form water-filled pores across the lipid bilayer of the outer membrane thus enabling hydrophilic molecules, of molecular weight up to 600–700, to move easily across the membrane. Larger water-soluble compounds may also use the pores provided they are sufficiently flexible to 'wriggle' through. In general the permeability afforded by the pores is non-specific. Certain water-soluble antibiotics move freely through the pores, and mutations which adversely affect the matrix proteins may decrease bacterial sensitivity to such drugs.

Porins have been identified in many Gram-negative bacteria including the potentially dangerous pathogen *Pseudomonas aeruginosa*. This organism is notorious for its intrinsic resistance to most antibacterial drugs. Experimental evidence indicates that the porins of *Pseudomonas aeruginosa* are able to form pores larger than those of *Escherichia coli*. In the living *Pseudomonas aeruginosa*, however, it is believed that most of the pores are usually closed and this may account in part for the striking impermeability of this organism to so many drugs. The physiological role of porins in providing routes of access for essential water-soluble molecules across the outer membrane has clear consequences for their utilization by antibacterial drugs. There is, for example, an inverse correlation between the hydrophobicity of β-lactam compounds and their rates of penetration across the outer membrane. Not surprisingly, the more hydrophobic compounds are less able to traverse water-filled pores. However, differences in rates of antibiotic diffusion across the outer membrane alone do not necessarily affect microbial susceptibility. For example, although the rate of penetration of the *Escherichia coli* outer membrane by carbenicillin is only 1/32 that of cephaloridine, the minimal inhibitory concentrations of these β-lactams are very similar. However, susceptibility can be adversely affected if slow penetration of the outer membrane is synergized by other factors, e.g. drug inactivation by enzymes in the deeper layers of the cell.

Porins exhibit certain general preferences for the type of solute permitted to traverse the pores. The porins of *Escherichia coli* accept cationic species in preference to anions which in turn diffuse more slowly than neutral molecules. The porins of *Neisseria gonorrheae* on the other hand prefer anions to cations. Zwitterions overall diffuse more rapidly than monoionic species.

Solute-specific channels

Specific channels or receptors exist in the outer membrane that facilitate the transport of nutrients such as vitamin B_{12}, mono- and oligo-saccharides,

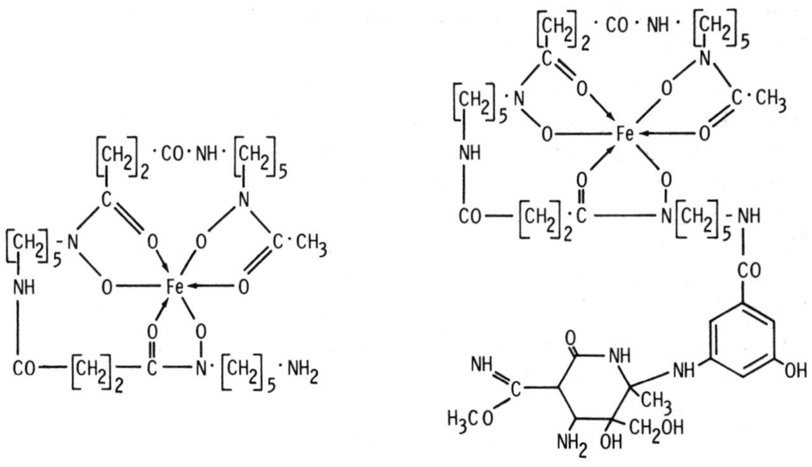

Figure 7.1 Ferrioxamine B, an iron-chelating growth factor, and ferrimycin A_1, a related antibiotic.

nucleosides, amino acids and various iron-transporting compounds such as sideramines and enterochelins. The mechanism of action of the receptor proteins is not understood but their function is linked in some way to that of the protein product of the *ton B* gene. Mutations in *ton B* indirectly negate the function of the receptor proteins. Antibiotics believed to depend upon iron-transporter receptor proteins for crossing the outer membrane include ferrimycin A_1 (Fig. 7.1) which closely resembles the iron-transporting growth factor ferrioxamine B and a recently described cephalosporin, E-0702 which bears an iron-chelating catechol moiety. The remarkable activity of the latter compound against the usually recalcitrant *Pseudonomas aeruginosa* is attributed to the subversion of the iron-transport system leading to high concentrations of the drug at the site of the β-lactam-binding proteins (PBPs, see Chapter 2).

The influence of the major structures of bacterial cell envelopes on the penetration of antibacterial agents is summarized in Table 7.2.

SOME EXAMPLES OF MODES OF PENETRATION OF ANTIMICROBIAL AGENTS

Passive diffusion

Drugs that probably cross the cytoplasmic membrane by passive diffusion include quinolones, trimethoprim, sulphonamides, macrolides, rifamycins

Table 7.2   Features of the bacterial cell envelope that influence the penetration of antibacterial agents

Structure	Effects on penetration
Lipopolysaccharides of Gram-negative outer membrane	Hinder access of hydrophobic antibiotics to the lipid bilayer
Lipid bilayer of Gram-negative outer membrane	Rate of penetration generally dependent on lipophilicity of solute, i.e. simple diffusion of water-soluble drugs is severely hindered
Hydrophilic pores of Gram-negative outer membrane	Permit rapid penetration through outer membrane of water-soluble molecules up to 650 daltons
Nutrient receptor proteins of Gram-negative outer membrane	Ensure penetration through outer membrane of molecules which bind to the receptor proteins
Teichoic and teichuronic acids of Gram-positive bacteria	Strongly anionic character of these polymers may affect rate of penetration of ionized molecules
Lipid bilayer of cytoplasmic membrane	Rate of passive diffusion generally dependent on lipophilicity of solute, thus penetration of water-soluble drugs is hindered
Nutrient-transport proteins of cytoplasmic membrane	Facilitate rapid penetration of agents chemically related to natural nutrients, e.g. COOH-substituted peptides, D-cycloserine, fosfomycin. May also enhance penetration of some structurally remote compounds like tetracyclines and chloramphenicol

and fusidic acid. However, experience has shown that it is not always easy to assign a drug exclusively to a particular mechanism of membrane transport. For example the means by which chloramphenicol crosses the cytoplasmic membrane has been debated for years. A consensus appeared to agree that passive diffusion provided an explanation until recent experiments with *Haemophilus influenzae* indicated that facilitated diffusion may after all play a significant role, at least in this important pathogen. It seems likely that in many cases more than one mode of transport contributes to intracellular penetration. Although passive diffusion of some drugs across the cytoplasmic membrane is slow, appreciable binding to their intracellular targets promotes a net inward flux until the binding sites are saturated. If the rate of occupation of the target sites is not substantially exceeded by the rate of synthesis of new sites, inhibition of cell growth and division soon follows.

Facilitated diffusion

D-CYCLOSERINE

This antibiotic, which is a structural analogue of D-alanine (Chapter 2), enters bacterial cells by the transport system in the cytoplasmic membrane for D-alanine. In *Streptococcus faecalis* D- and L-alanine use the same transport mechanism and D-cycloserine competitively inhibits the uptake of both isomers. In *Escherichia coli*, however, D- and L-alanine are transported separately. In addition there are high- and low-affinity transport systems for D-alanine. At the minimal inhibitory concentration of D-cycloserine $(4\mu\mathrm{M})$ for *Escherichia coli*, the drug favours the high-affinity D-alanine transport system. The facilitated transport system for D-alanine is energy-coupled via the proton motive force and results in the intracellular accumulation of D-alanine and D-cycloserine against their concentration gradients. This explains why in order to produce significant inhibition of L-alanine racemase and D-alanyl-D-alanine synthetase in cell-free preparations, the concentration of D-cycloserine must be considerably higher than the minimal growth-inhibiting concentration.

TETRACYCLINES

The discovery that the tetracyclines are actively transported into bacterial cells but not into mammalian cells helped to explain the antibacterial specificity and general lack of toxicity of these drugs (Chapter 5). The lack of any obvious structural similarity between the tetracyclines and known transportable cell nutrients has precluded a straightforward explanation of how the drugs are taken up by bacteria. In Gram-negative bacteria the diffusion of the more hydrophilic tetracyclines across the outer membrane is porin mediated. The lipophilic derivative minocycline, however, probably diffuses through the lipid phase rather than through the pores. The events in the cytoplasmic membrane that bring about the marked intracellular accumulation of the tetracyclines are more obscure. Accumulation is partially coupled to the energy metabolism of the cells and may involve both the proton motive force and the hydrolysis of ATP. Although tetracyclines probably cross the cytoplasmic membrane with the aid of protein carriers, direct evidence for this is lacking. The penetration of the cytoplasmic membrane by minocycline is probably at least partly by passive diffusion. As we shall see in Chapter 8, reversal of the active accumulation of tetracyclines underlies the commonest form of resistance to these antibiotics.

SIDEROMYCINS

These complex antibiotics, e.g. ferrioxamine B (Fig. 7.1), whose mode of action is uncertain subvert the bacterial transport of sideramines which is normally concerned with the uptake of iron. Sideromycins traverse the outer membrane following their interaction with the specific sideramine-receptor protein. Transport of the sideramine–iron complexes across the cytoplasmic membrane involves facilitated diffusion coupled to energy metabolism. The sideromycin antibiotics also use this system to gain access to the cytoplasm. The maximum intracellular concentration of the sideromycin, A22765, is about one hundredfold greater than that in the medium. The antibiotic is, however, extensively bound to intracellular sites, i.e. the apparent concentration may not be all 'free' drug. Mutations affecting the outer membrane receptors for sideramines and the cytoplasmic membrane transport mechanism both give rise to resistance to the sideromycins.

AMINOGLYCOSIDES

These ionized water-soluble molecules are believed to penetrate the outer membrane of Gram-negative bacteria via the porin system. In addition, aminoglycosides may enhance their penetration by disorganizing the outer membrane as a result of the competitive displacement of essential cations from the lipopolysaccharide. The initial uptake of dihydrostreptomycin by *Escherichia coli* is characterized by a rapid electrostatic interaction between the positively charged guanidino groups of the antibiotic and anionic residues in the outer layers of the bacterium. This is followed by a slow energy-dependent penetration into the bacterial cytoplasm. After 15–30 min a third phase of uptake begins in which there is rapid intracellular accumulation of dihydrostreptomycin also linked to energy processes in the cell but perhaps not directly to respiration. In *Escherichia coli* at least, this final phase appears to be irreversible; the antibiotic can only be released from the cells by fatally damaging the cell membranes with organic solvents such as toluene. The molecular mechanisms involved in the energy-dependent phases of aminoglycoside uptake are obscure. An earlier suggestion that an aminoglycoside-inducible permease, normally responsible for accumulating polyamines, is involved has now been discounted. The intrinsic resistance of anaerobic bacteria to the aminoglycosides may be due to their limited ability to accumulate these antibiotics. A recent hypothesis suggests that aberrant proteins synthesized under the influence of low concentrations of aminoglycosides become integrated into the membrane rendering it more permeable to antibiotics.

FOSFOMYCIN (PHOSPHONOMYCIN)

This simple phosphorus-containing antibiotic uses two different physiological transport systems to gain access to the bacterial cytoplasm. (i) The permease for α-glycerophosphate is used by fosfomycin which it structurally resembles. The ability of fosfomycin to use this permease is competitively antagonized by high concentrations of α-glycerophosphate. (ii) The permease for hexose 6-phosphates in certain Enterobacteriaceae and staphylococci is also exploited by fosfomycin. Both transport systems are induced by their physiological substrates, although not by fosfomycin. In an experimental infection with *Escherichia coli* in mice, the therapeutic efficacy of fosfomycin was enhanced by pretreatment of the animals with glucose 6-phosphate. This can be attributed to the induction of the permease in the infecting cells permitting a higher intracellular concentration of fosfomycin. In Gram-negative bacteria fosfomycin crosses the outer membrane by pores used by α-glycerophosphate and the hexose 6-phosphates.

THE EXPLOITATION OF TRANSPORT SYSTEMS IN THE DESIGN OF NEW ANTIMICROBIAL AGENTS

The barrier function of the microbial cell provides both problems and opportunities for the designer of new drugs. In Gram-negative cells the outer membrane will be more easily breached by compounds that can utilize the pore system or the specific nutrient receptors. Further study of the unusual permeability characteristics of the outer envelope of *Pseudomonas aeruginosa* may assist the design of new agents to combat this troublesome pathogen.

The discovery of relatively non-specific permeases for small peptides in bacteria and other micro-organisms led to the demonstration that poorly permeating toxic amino acid analogues have enhanced antibacterial activity when attached to small peptides. The cytoplasmic membrane of *Escherichia coli* possesses numerous permeases that facilitate the acquisition of small peptides, of up to six or seven amino acids in length. A free *N*-terminus is essential for peptide transport, but a variety of carboxyl-substituted peptides are accepted by the permeases. Small toxic substituents at the *C*-terminus can therefore be 'smuggled' into the cell. An interesting example of a synthetic peptide analogue that exploits peptide permeases is alafosfalin (Fig. 7.2). This compound, in which the carboxyl residue of alanylalanine is replaced by a phosphonic acid group, is rapidly accumulated by *Escherichia coli* to an intracellular concentration 100–1000 times that in the medium. A

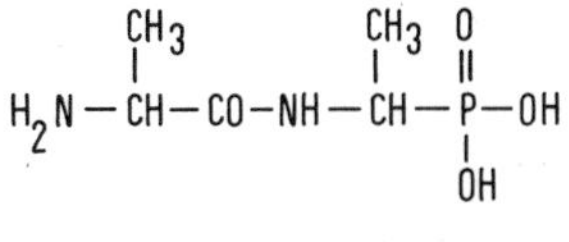

$$\underset{\text{H}_2\text{N}}{} - \text{CH} - \text{CO} - \text{NH} - \text{CH} - \overset{\overset{\text{O}}{\|}}{\text{P}} - \text{OH}$$

Figure 7.2 Alafosfalin or L-alanyl-L-1-aminoethylphosphonic acid, a new antibacterial drug that exploits the peptide permease systems of bacterial cells to gain access to its intracellular target enzyme.

tripeptide permease is the main agent for the uptake of alafosfalin and requires L–L stereochemistry in the dipeptide. Inside the cell, alafosfalin is cleaved by aminopeptidase action to release L-1-aminoethylphosphonic acid which inhibits alanine racemase and consequently the biosynthesis of the peptidoglycan of the bacterial wall (cf. D-cycloserine, Chapter 2). Although alafosfalin has shown activity against infections in man, it has not established itself in the therapeutic armoury perhaps because of its limited antibacterial spectrum and the ease with which bacteria develop resistance to it by the deletion of non-essential oligopeptide permeases.

Fungal infections are frequently refractory to conventional antibiotic therapy; in some instances this is due to the limited penetration of antibiotics into the cells. However, synergism has been observed between a number of antifungal agents and the polyene, amphotericin B. The polyene may assist the penetration of impermeant inhibitors by disorganizing the cytoplasmic membrane. Facilitated transport of peptides also occurs in fungi and it is possible that this could be exploited to introduce inhibitory molecules into fungal cells. The naturally occurring antifungal antibiotics, polyoxin D and nikkomycin (Chapter 6) use peptide permeases to penetrate sensitive fungal cells. In this instance nature appears to be a step ahead of the medicinal chemist.

FURTHER READING

Chopra, I. and Ball, P. (1982) Transport of antibiotics into bacteria. *Adv. Microb. Physiol.*, **23**, 183.

Ferro-Luzzi Ames, G. (1986) Bacterial periplasmic transport systems: structure, mechanism and evolution. *Annu. Rev. Biochem.*, **55**, 397.

Hancock, R. E. W. (1984) Alterations in outer membrane permeability. *Annu. Rev. Microbiol.*, **38**, 237.

Nikaido, H. and Vaara, M. (1985) Molecular basis of bacterial outer membrane permeability. *Microbiol. Rev.*, **49**, 1.

Ringrose, P. S. (1985) Warhead delivery and suicide substrates as concepts in antimicrobial drug design. *The Scientific Basis of Antimicrobial Therapy, Society of General Microbiology Symposium 38*, (eds. D. Greenwood and F. O'Grady).

Scarborough, G. A. (1985) Binding energy, conformational change and the mechanism of transmembrane solute movements. *Microbiol. Rev.*, **49**, 214.

Tute, M. S. (1971) Principles and practice of Hansch analysis, a guide to structure–activity relationships for the medicinal chemist. *Adv. Drug Res.*, **6**, 1.

8 Resistance to antimicrobial drugs

The development of safe effective antimicrobial drugs has revolutionized medicine in the last fifty years. Morbidity and mortality from microbial disease have been drastically reduced by modern chemotherapy. Unfortunately, micro-organisms are nothing if not versatile, and the brilliance of the chemotherapeutic achievement has been somewhat dimmed by the emergence of microbial strains presenting a formidable array of defences against our most valuable drugs. This should not surprise us, since the evolutionary history of living organisms is concerned with their adaptation to the environment. The adaptation of micro-organisms to the toxic hazards of antimicrobial drugs is therefore probably inevitable.

The first detailed account of microbial drug resistance was given by Paul Ehrlich in 1907 when he encountered this problem soon after the development of arsenical chemotherapy against trypanosomiasis. As the sulphonamides and antibiotics were brought into medical and veterinary practice resistance against these agents began to emerge. Microbial resistance to drugs is now widespread and constitutes an ever-present threat to the successful treatment of infectious disease.

Our intention in this chapter is to give an outline of the genetic background to the problem of drug resistance and then to describe the general types of biochemical mechanisms that give rise to resistance. We shall draw upon a number of important examples of drug resistance to illustrate these underlying biochemical mechanisms.

THE GENETICS OF DRUG RESISTANCE

The tremendous advances made in the science of bacterial genetics over the past 40 years have found a most important practical application in furthering our understanding of the problem of drug resistance. As a result we now have a fairly complete picture of the genetic factors underlying the emergence of drug-resistant bacterial populations. In contrast, the study of the genetics of resistance in pathogenic fungi and protozoa is much less advanced and the

173

lack of knowledge hinders a proper appreciation of the spread of resistance amongst these organisms.

The early studies on the genetics of drug resistance were bedevilled by an exhausting controversy. On the one hand were those who believed that the development of a resistant cell population could be explained by the phenotypic adaptation of the cells to an inhibitory compound without significant modification in their genotype. The opposing faction took the view that any large population of cells which was sensitive overall to a drug was likely to contain a few genotypically resistant cells. The continued presence of the drug resulted in the emergence of a new population of resistant cells by a process of selection.

Evidence gathered over the years strongly supports the second of these two theories. As we shall see, there are examples of phenotypic adaptation of bacterial cells to growth-inhibitory drugs; however, such cells are always genotypically different from the sensitive cells and generally do not constitute the majority of a wild-type population not previously exposed to the drug.

When the selective pressure applied by an antimicrobial drug is removed, a previously resistant microbial population may revert to drug sensitivity. Sometimes the resistant cells are at a selective disadvantage to drug-sensitive cells in a drug-free environment and are therefore eventually outgrown by the sensitive cells. In other examples of reversion, the genetic material that confers drug resistance is lost from the cells and the drug-resistant cells are progressively diluted out of the population.

Spontaneous mutations

Gene mutations are said to be spontaneous when they occur unprovoked by experimental mutagenic procedures. Spontaneous mutations are relatively infrequent, e.g. one mutation per 10^5-10^7 cells per cell division. However, when the vast numbers of cells involved in bacterial populations are considered, the probability of a mutation causing increased drug resistance is quite high. The simple and elegant technique of replica plating gives a convincing demonstration that spontaneous mutations to drug resistance regularly occur in sensitive populations in drug-free media (Fig. 8.1). A spontaneous mutation may occasionally cause a large increase in resistance but more usually resistance develops as a result of numerous mutations each giving rise to a small increment in resistance. In the second case highly resistant cells emerge only after prolonged or repeated exposure of the cell population to the drug.

At one time it was believed that spontaneous mutations to drug resistance, followed by the selection of resistant cells in the presence of drug, explained the emergence of drug-resistant populations. However, the realization that bacteria can acquire additional genetic material by conjugation, transforma-

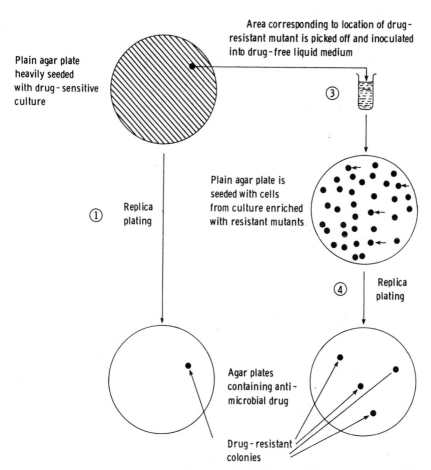

Figure 8.1 The technique of replica plating reveals the existence of drug-resistant cells in a population that is overall drug sensitive. A plain agar plate is heavily seeded with cells from the drug-sensitive culture, and is incubated until growth occurs. Cells are transferred by a velvet pad to a plate containing the antibacterial drug; this plate is then incubated and the position of any colonies noted. The area on the drug-free plate corresponding to the location of the resistant colony on the drug plate is picked off and cultured in drug-free medium. Although still contaminated with sensitive cells, this culture will contain many more resistant cells than the original culture. Plating out of the 'enriched' culture on a plain plate followed by replication to a drug plate therefore reveals a higher incidence of drug-resistant colonies. The experiment indicates that drug-resistant mutants occur in a bacterial population not previously exposed to the drug.

tion and transduction led to the conclusion that spontaneous mutations make only a minor contribution to the clinical problem of drug resistance.

SPREAD OF DRUG RESISTANCE BY GENE TRANSFER

Conjugation

GRAM-NEGATIVE BACTERIA

The spread of drug resistance by cellular conjugation in Gram-negative bacteria harboured by animals and man is recognized as an important clinical threat to the treatment of diseases caused by these organisms. The transfer of genes by cell conjugation in Gram-negative bacteria was discovered some years before its importance in drug resistance was appreciated.

The realization that drug resistance could be transferred by cell conjugation came from combined epidemiological and bacterial genetic studies in Japan. The first clue was provided by the isolation, from patients suffering from dysentery, of strains of *Shigella* that were resistant to a number of drugs, including sulphonamides, streptomycin, chloramphenicol and tetracycline. Even more striking was the discovery that both sensitive and multiresistant strains of *Shigella* could occasionally be isolated from the same patient in the same epidemic. Most patients harbouring multiresistant *Shigella* also had multiresistant *Escherichia coli* in the intestinal tract. This suggested that the drug-resistance markers might be transferred from *Escherichia coli* to *Shigella* and vice versa. Subsequently it was confirmed that Gram-negative bacteria can indeed transfer drug resistance not only to cells of the same species but also to bacteria of different species and even different genera.

The genetic elements associated with the transference of drug-resistance markers by cellular conjugation are called R-plasmids (Fig. 8.2). Under most conditions they are separate from the bacterial chromosome. They consist of two distinct but frequently linked entities: (a) the genes which initiate and control the conjugation process; (b) a series of one or more linked genes that confer resistance to specific antibacterial agents. The conjugative region is analogous to the F-plasmid which also confers on Gram-negative bacteria the ability to conjugate with cells lacking an F-plasmid. The complete R-plasmid resembles the F-prime plasmid (F′) in carrying genetic material additional to that which controls conjugation.

A bewildering variety of R-plasmids has been described carrying various combinations of drug-resistance determinants. Apart from the obvious property of drug resistance, the other phenotypic characteristics conferred on the cells by different R-plasmids have prompted attempts at classification. The characteristics include: (i) the ability (fi$^+$) or inability (fi$^-$) to repress the fertility properties of an F-plasmid present in the same cell; (ii) the type of

Resistance
plasmid

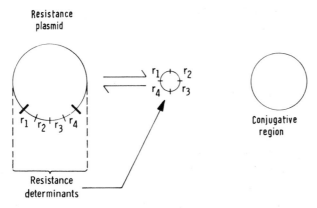

Conjugative
region

Resistance
determinants

Figure 8.2 Diagrammatic representation of an R-plasmid and its reversible dissociation into two components. Although not shown in the diagram, if the R-plasmid
carries a determinant for tetracycline resistance this determinant almost certainly
remains linked to the R-plasmid after dissociation whereas the other resistance
determinants $r_1, r_2, r_3, r_4 \ldots r_n$ form a separate replicon.

sex pilus (see below) that the R-plasmid determines; (iii) the inability of an
R-plasmid to co-exist in a bacterium with certain other plasmids leading to
the division of R-plasmids into incompatibility groups; (iv) the presence of
genes in the R-plasmid that specify DNA restriction and modification
enzymes. An R-plasmid cannot be defined on the basis of a single characteristic and a combination of properties must be used to classify it.

Transposons

Advances in molecular genetics provide a likely explanation for the very wide
distribution of drug-resistance determinants among the plasmids and
chromosomes of many bacterial species. For many years the movement of
genes among plasmids and chromosomes was believed to result from classical recombination dependent on the product of the bacterial *recA* gene and
the reciprocal exchange of DNA in regions of considerable genetic homology.
This permits the exchange of genetic information only between closely
related genomes. Such a restricted phenomenon seemed unlikely to explain
the widespread distribution of specific resistance determinants. It is now
clear that the acquisition of genetic material by plasmids and chromosomes
in both Gram-negative and Gram-positive bacteria is not limited by the
classical *recA*-dependent recombination. Certain replicons, known as transposons, are able to insert themselves into a variety of genomic sites that have
no common ancestry, i.e. homology, with the inserting sequence. In the
simplest transposons the whole of the genetic information is concerned with
the insertion function. More complex transposons contain additional genes

that determine a variety of functions including resistance to antibiotics. Transposons have been found with resistance determinants for ampicillin, tetracycline, kanamycin, streptomycin, sulphonamides, chloramphenicol and trimethoprim. The biological significance of these complex transposons is that they facilitate the movement of resistance determinants between chromosomes, R-plasmids and phages. The widespread use of antibacterial drugs in medical and veterinary practice and in animal husbandry has encouraged the selection of bacterial plasmids with transposons. Almost certainly the traffic of transposons independent of *recA* has produced the accumulation of replicons with sets of genes conferring resistance to numerous antibiotics.

Molecular properties of R-plasmids

R-plasmids can be isolated from host bacteria as circular DNA (Fig. 8.3) in both closed and 'nicked' forms. There has been some debate as to the proportions of closed and nicked circles that arise during the isolation procedures. Nevertheless there is little doubt that both forms exist in the cell; the closed circular structure is probably adopted by R-plasmids when not engaged in replication. The contour lengths and thus molecular weights of isolated R-plasmids depend very much on the host bacterium and upon the culture conditions prevailing immediately before the isolation procedure. The R-plasmid may sometimes dissociate into its conjugative and resistance (r) determinants. This is more common in some host species, e.g. *Proteus mirabilis* and *Salmonella typhimurium*, than in *Escherichia coli*, where dissociation hardly ever occurs. The dissociation event seems to depend on the activity of a simple transposon that may be inserted at the junction of the two regions. Estimates of the molecular weights of conjugative regions from different R-plasmids lie between 50×10^6 and 60×10^6, with a buoyant density of 1.709. The molecular weight of the resistance determinant depends on the number of drug-resistance genes carried, but typically it is much lower than that of the conjugative region. Thus a determinant carrying markers for resistance to chloramphenicol, streptomycin, spectinomycin and sulphonamide has a molecular weight of 12×10^6 with a buoyant density of 1.717.

The numbers of R-plasmids harboured by individual bacteria is influenced by the properties of the plasmid and host as well as by the culture conditions. As a general rule (to which there are exceptions), the larger R-plasmids are present only in a limited number of copies (one to four) per chromosome in *Escherichia coli*, whereas in *Proteus mirabilis* the number is much more variable and varies during the growth cycle. In stationary phase, replication of the R-plasmid continues for some time after chromosomal replication has ceased. Conditions that give rise to an increased number of R-plasmid copies are sometimes associated with enhanced resistance. However, the level of

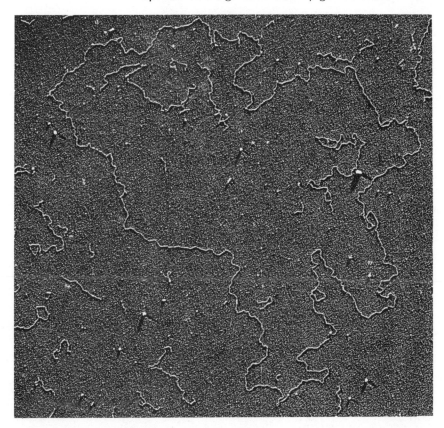

Figure 8.3 Electron micrograph of R-plasmid DNA isolated from *Proteus mirabilis* harbouring an R-plasmid with resistance markers to streptomycin, sulphonamides and chloramphenicol. The circular DNA molecule has a total length of 28.5 μm. This photograph is reproduced by kind permission of Dr Royston Clowes and the American Society for Microbiology (*J. Bacteriol.*, **97**, 383 (1969)).

resistance does not always reflect the number of resistance gene copies. For example, although the number of R-plasmid copies is frequently greater in *Proteus mirabilis* than in *Escherichia coli* the level of drug resistance expressed in the former organism is usually lower than in *Escherichia coli*. This is the case for resistance to several different antibiotics, and factors other than the number of gene copies may be more important in determining drug resistance. Resistance-gene-dosage effects are not known to be significant in the treatment of bacterial infections.

The conjugation process

Cells bearing an R-plasmid (R^+) are characterized by the ability to produce surface appendages known as sex pili. The sex pili of R^+ bacteria closely resemble those produced by F^+ organisms. When 'male' R^+ cells are mixed with sensitive R^- 'female' cells, mating pairs are immediately formed by some sort of surface interaction possibly involving the sex pili. The transfer of a copy of the R-plasmid from the R^+ to the R^- cell begins immediately and the acquisition of the R-plasmid by the 'female' cell converts it to a fertile drug-resistant cell that can in turn conjugate with other R^- cells. In this way drug resistance spreads rapidly through the bacterial population. Many of the fine details of the conjugation processes are still lacking; for example the role of the pilus in the transfer of plasmids is uncertain. Despite the attraction of the idea that the plasmid passes through the bore of the pilus, it is very difficult to prove directly. The significance of the replication of the R-plasmid DNA to the transfer process is also not clear. The R-plasmid copy is very likely transferred as a linear single-stranded molecule which is converted to a duplex immediately it gains access to a replication site in the recipient cell. Fortunately perhaps, the frequency of R-plasmid transfer is much lower than that of F transfer. After infection of an R^- cell with an R-plasmid a repressor substance is thought to accumulate which eventually inhibits sex pilus formation. The ability to conjugate is therefore restricted to a short period immediately after acquisition of the R-plasmid. Sex pilus production in F^+ cells, by contrast, is not under repressor control and conjugatory activity is therefore completely unrestricted. Mutant R-plasmids exist which have lost the ability to restrict sex pilus formation and the frequency of R-plasmid transfer among bacteria bearing this type of plasmid is much higher.

Clinical importance of R-plasmids

It is generally agreed that R-plasmids existed before the development of modern antibacterial drugs. Clearly though, the widespread use and abuse of these drugs have led to a vast increase in the incidence of drug resistance caused by R-plasmids. This has been especially noticeable in farm animals which in many countries receive clinically valuable antibacterial drugs in their foodstuffs as growth promoters. These animals act as a reservoir for Gram-negative bacteria, such as *Escherichia coli* and *Salmonella typhimurium*, harbouring R-plasmids potentially transferable to man. Fortunately, some countries have restricted the use of growth promoters which are valuable drugs in human clinical medicine. This has led to a significant decline in the incidence in farm animals of R-plasmids mediating resistance to clinically important drugs.

The adverse contribution of drug resistance caused by R-plasmids to human morbidity and mortality is undeniable. For example, in neonatal diarrhoea caused by certain pathogenic strains of *Escherichia coli*, although the major requirement in the treatment of this distressing disease is the prevention of fatal dehydration, elimination of the pathogenic organisms may also be important and this is often difficult in the event of multiple resistance to the commonly used antibacterial agents. In one notorious outbreak the children were infected with a pathogenic strain of *Escherichia coli* resistant to penicillin, streptomycin, chloramphenicol, tetracycline, erythromycin, neomycin, novobiocin, cloxacillin and ampicillin. The infection eventually responded to gentamicin which was the only drug of those tested to which the pathogenic bacteria were sensitive. Another potentially alarming development has been the occasional appearance of *Salmonella typhi*, the causative organism of typhoid, with an R-plasmid conferring resistance to chloramphenicol and cotrimoxazole, the drugs most commonly used to treat this disease.

Ecological features fortunately tend to limit the clinical threat posed by R-plasmids. Firstly the environment of the gastrointestinal tract almost certainly depresses the conjugal activity of R^+ bacteria below that observed under ideal culture conditions in the laboratory. The emergence of an R^+ population of bacteria during antibiotic therapy is more likely to result from selection of resistant cells than from extensive conjugal transfer of resistance. After cessation of antibiotic treatment, the number of R^+ bacteria in the faeces declines; it seems that in the absence of drugs the colonizing ability of R^+ *Escherichia coli* in the gut is lower than that of the normal R^- residents. R-plasmids do not themselves impair the colonizing ability of bacteria, but the organisms that preferentially harbour them tend to be poor colonizers.

GRAM-POSITIVE BACTERIA

Transfer of drug-resistance genes by conjugation is not confined to Gram-negative bacteria. Conjugal transference of genes has now been described in various species of *Streptococcus*, *Staphylococcus*, *Bacillus* and *Clostridium*. Genes coding for conjugation occur on plasmids, prophages and transposons. In the *Streptococci* there are two groups of plasmids, coding for high (10^{-3}–10^{-1} per donor) and low ($<10^{-6}$ per donor) frequency of gene transfer. Donor cells sense the presence of recipients through the secretion by the latter of short pheromone peptides. The donors are stimulated to produce a protein called adhesin which induces aggregation of donor and recipient cells. Transference of resistance genes takes place within these cellular aggregates. Conjugation and gene transfer also occur in *Streptococci* apparently devoid of plasmid DNA. The information encoding these processes is found on elements called conjugative transposons which insert into several sites in the chromosome. These transposons can also insert into plasmids but generally seem to prefer a chromosomal location. Although conjugative transposons have so far only been clearly identified in the genus *Streptococcus*, it seems possible that similar genetic elements may mediate the transfer of resistance genes in *Clostridium difficile* and *Bacillus fragilis*. Despite these mechanisms for transferring resistance in *Streptococcus*, acquired drug resistance has not been a major clinical problem in these organisms which are usually treated with β-lactams.

In recent years the incidence of hospital-associated infections caused by *Staphylococcus aureus* resistant to many antimicrobials has risen alarmingly. Phage-mediated conjugation amongst *Staphylococcus aureus* and interspecific conjugation with *Staphylococcus epidermidis* may play a role in this damaging spread of drug resistance. The mechanism of conjugal gene transfer in *Staphylococcus* (approximately 10^{-6} per donor) is not yet understood but it is stimulated by dry and absorbent surfaces such as wound dressings, clothing and bedding materials.

Transduction

During the process of phage transduction, which occurs in both Gram-positive and Gram-negative bacteria, genetic information is transferred by phage particles from one bacterial cell to a related phage-susceptible cell. In lysogeny the DNA of the infecting temperate phage is inserted into the bacterial genome where it replicates synchronously with the bacterial DNA until the integrated temperate phage reproduces itself virulently either spontaneously or as a result of treatment with ultraviolet light or certain chemicals. Complete phage particles are produced and the bacterial cell is lysed. However, before unrestricted replication of the phage DNA occurs it is first released from the bacterial chromosome. During this phase short stretches of the bacterial chromosome that were contiguous with the phage DNA in the integrated state may be carried with the phage DNA and subsequently replicated as part of the phage genome. A similar process can result in the acquisition by the phage of genetic material from plasmids. The reader is referred to a suitable text on bacterial genetics for a fuller discussion of the complexities of phage transduction. Nevertheless, it can be seen that a transducing phage might pick up drug-resistance determinants from a resistant bacterium during the process of induction either from the chromosome or from a plasmid. Lysis of the cell releases multiple copies of the phage each carrying a copy of the drug-resistance marker. Subsequent infection of phage-susceptible drug-sensitive cells under conditions favouring lysogeny results in the newly infected cells becoming drug resistant. In some cases a drug-resistance determinant from the original host is incorporated into the phage particle only at the expense of part of the phage genome. The resulting defective phage, while conferring the property of drug resistance on infected recipient cells, cannot enter the lysogenic state, i.e. it cannot be subsequently induced to lyse the cells. However, provided that the phage DNA, together with the acquired resistance marker replicates normally in the recipient cells, a clone of drug-resistant cells arises from each parent infected with a transducing phage. The transduction of plasmid determinants of drug resistance usually occurs with much greater facility than that of chromosomal determinants. Possibly this reflects the ability of plasmids to replicate in a variety of environments, whereas the replication of chromosomal DNA is closely adapted to a specific organism.

The transfer of drug-resistance markers by transduction in *Staphylococcus aureus* can easily be demonstrated *in vitro* although whether it occurs to any significant extent *in vivo* is much less certain.

Transformation

Drug resistance can be transferred *in vitro* to sensitive bacteria by treating them with DNA extracted from drug-resistant mutants. Since transformation of the morphological characteristics of pneumococci from smooth to rough forms is believed to occur *in vivo* in infected animals by this process, it is conceivable that transformation of sensitive cells to drug resistance could also occur *in vivo*. However, it should be remembered that transformation is a relatively inefficient way of spreading genetic information, since for each resistance marker acquired by a sensitive cell, a resistant cell must lyse in order to release its DNA. Genetic transformation, therefore, probably does not contribute substantially to the clinical problem of drug resistance.

BIOCHEMICAL MECHANISMS OF DRUG RESISTANCE

Some years ago the American microbiologist Bernard Davis listed a number of possible mechanisms for drug resistance. While this list continues to be useful, experience has shown that mechanisms 1–4 (below) are the most common, at least in bacteria.

Summary of possible mechanisms

1. Conversion of active drug to an inactive derivative by enzyme(s) produced by the resistant cells
2. Modification of the drug-sensitive site
3. Loss of cell permeability to a drug
4. Synthesis of an additional drug-resistant enzyme or overproduction of a drug-sensitive enzyme
5. Increased concentration of a metabolite that antagonizes the inhibitor

We shall illustrate these various mechanisms with appropriate examples and indicate their relative importance to the general problem of microbial resistance. However, it should be realized that the resistance of bacterial cells to an antibiotic may not depend on just one mechanism. Especially in Gram-negative bacteria, resistance results both from the natural permeability barrier of the outer envelope that slows drug access and from more specific mechanisms of resistance that may inactivate a drug, prevent its transport across the cytoplasmic membrane or its binding to intracellular sites.

(1) Conversion of an active drug to an inactive derivative

β-LACTAMS

The destruction of penicillin by bacteria that produce penicillinase (β-lactamase) involves the opening of the β-lactam bond of penicillin to give the inactive penicilloic acid (Fig. 8.4). As we shall see later, the nature of the side chain, R, can be of critical importance in determining the susceptibility of the β-lactam bond to penicillinase. The related antibiotics, the cephalosporins,

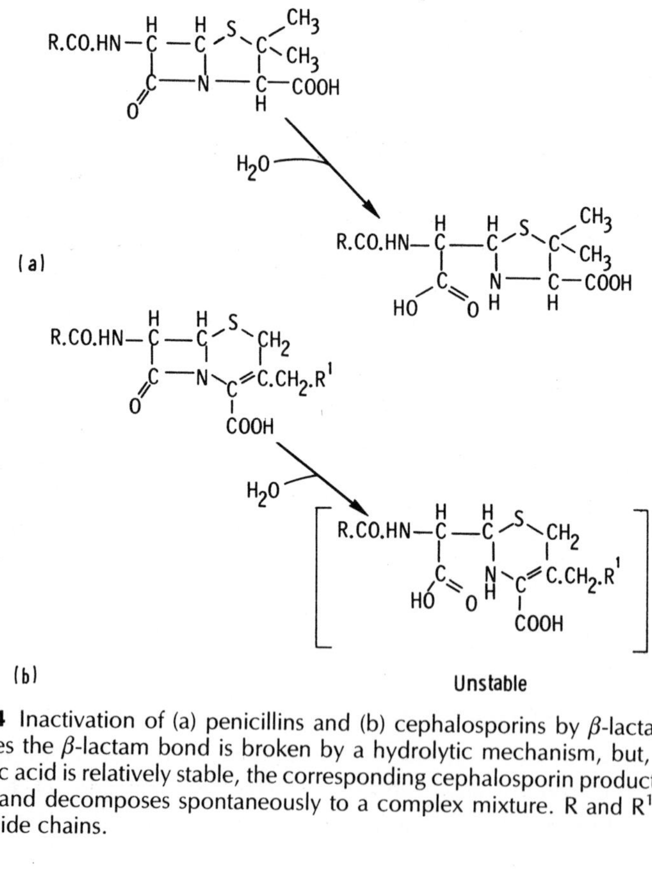

Figure 8.4 Inactivation of (a) penicillins and (b) cephalosporins by β-lactamase. In both cases the β-lactam bond is broken by a hydrolytic mechanism, but, whereas penicilloic acid is relatively stable, the corresponding cephalosporin product is highly unstable and decomposes spontaneously to a complex mixture. R and R^1 indicate variable side chains.

are also degraded by β-lactamase (Fig. 8.4). The immediate end product of cephalosporin degradation is cephalosporanoic acid, although this compound is so unstable that it spontaneously undergoes a complex series of degradative reactions. Penicillin and cephalosporin β-lactamases are produced by many Gram-positive and Gram-negative bacteria. Individual enzymes exhibit preferences for either penicillins or cephalosporins as substrates, so that cross-resistance between the two classes of β-lactam antibiotics is frequently incomplete.

Gram-positive β-lactamases

The most important β-lactamase in Gram-positive bacteria is that produced by *Staphylococcus aureus* which was responsible for the alarming rise in the resistance of this pathogen to penicillin that occurred in the late 1940s and 1950s. The β-lactamase of *Staphylococcus aureus*, of which there are four minor variants, is an inducible enzyme. Enzyme production is very low in the absence of penicillin or cephalosporin. When the β-lactamase gene is de-repressed by the addition of minute quantities of antibiotic (as little as 0.0024 μg ml^{-1} of medium) enzyme production is greatly increased and may account for more than 3% of the total protein being synthesized by the bacterium. The enzyme, which is predominantly a penicillinase, is released from the bacterial cell and inactivates the antibiotic in the surrounding medium. Considerable dilution of β-lactamase occurs and this is the source of the observed 'inoculum effect'. In a small inoculum the few cells present cannot destroy all the surrounding antibiotic, but in an inoculum containing many cells the much larger production and release of β-lactamase can overcome the challenge. Staphylococcal resistance to penicillin is therefore very dependent upon inoculum size.

The β-lactamase of *Staphylococcus aureus*, which has a molecular weight of approximately 29000, is related in its amino acid sequence to the inducible β-lactamases of *Bacillus licheniformis* and *Bacillus cereus*. In *Staphylococcus aureus* the gene for β-lactamase usually occurs on a plasmid which may also carry markers for resistance to other antibiotics. The specific problem posed by staphylococcal β-lactamase was thought to have been largely overcome by the development of novel semi-synthetic penicillins and cephalosporins that are poor substrates for β-lactamase and yet retain excellent antibiotic activity (see below). However, as we shall see, resistance to these β-lactamase-stable drugs can emerge in other ways.

Gram-negative β-lactamases

The nature of Gram-negative resistance to β-lactam antibiotics differs considerably from that of Gram-positive bacteria. The complex outer envelope of Gram-negative cells makes them intrinsically less sensitive to the original β-lactam antibiotics. However, soon after the appearance of semi-synthetic broad-spectrum compounds, such as ampicillin, resistance in Gram-negative pathogens due to β-lactamase activity began to emerge. Since those early days an extraordinary number of β-lactamases has been discovered in Gram-negative bacteria, so many in fact that it has proved difficult to devise ways of classifying them. The Gram-negative β-lactamases fall broadly into two groups: (1) enzymes determined by chromosomal genes – in many genera these enzymes are inducible by β-lactams but to a lesser extent than in Gram-positive bacteria; (2) enzymes determined by plasmid genes (of which there are at least 20) and produced constitutively in small amounts. Gram-negative β-lactamases remain largely associated with the cells and therefore hinder the access of β-lactams to the membrane-associated target sites by destroying the antibiotics within the cell envelope, probably in the periplasmic space.

The Gram-negative β-lactamases can be further subdivided on the basis of their activities against penicillins, cephalosporins and the many semi-synthetic variants of these molecules. Most chromosomally mediated β-lactamases inactivate both cephalosporins and penicillins, but are considerably more active against the former group. Chromosomal enzymes specific for penicillins are exceedingly rare. By far the most important plasmid-mediated β-lactamase is the TEM-1 enzyme which has a broad activity range against penicillins and cephalosporins. The wide distribution of TEM-1 almost certainly results from the gene for this enzyme being carried by transposons. The closely related enzyme, TEM-2, is less widely distributed than TEM-1.

Chemical approaches to the β-lactamase problem

The advent of the semi-synthetic β-lactams during the 1950s offered an escape from the problem of bacterial resistance caused by β-lactamase. Compounds such as methicillin and cloxacillin (Chapter 2) which have bulky substituents in the penicillin side chain were found to be poor substrates for β-lactamase. The affinity of methicillin for staphylococcal β-lactamase is much lower than that of benzylpenicillin, and the maximum rate of hydrolysis of methicillin by this enzyme is only one-thirtieth of that of benzylpenicillin. Until recently methicillin was an effective drug against infections caused by β-lactamase-producing staphylococci, even though its intrinsic antibacterial activity is substantially lower than that of benzylpenicillin. Methicillin is only slowly degraded by Gram-negative β-lactamases but is ineffective against Gram-negative infections because it fails to penetrate the outer envelope. To combat the menace of Gram-negative β-lactamases, therefore, compounds were needed that both resisted β-

lactamase attack and penetrated to the target sites in the cytoplasmic membrane. The substituted cephalosporins undoubtedly go a long way towards satisfying these requirements. Compounds such as cefuroxime, cefoxitin and cefotaxime (Chapter 2) combine good intrinsic activity against Gram-negative pathogens and considerable stability to the major β-lactamases.

Unfortunately a new mechanism of resistance to the so-called 'third generation' cephalosporins like cefotaxime has emerged in Gram-negative bacteria with chromosomally determined inducible β-lactamases. Clinical isolates of *Enterobacter cloacae*, *Pseudomonas aeruginosa*, *Proteus* spp. and *Citrobacter* spp. are increasingly found to produce β-lactamases constitutively as a result of a mutation affecting the normal repressor control. The large quantity of enzyme produced by these mutants (1–2% of total protein in some organisms) seems to outweigh their relatively poor activity against the cephalosporins and reduces the drug concentration in the periplasmic space below that needed for inactivation of the target penicillin-binding proteins. Even more stable cephalosporin molecules are therefore needed to cope with these dangerous new mutants.

The use of inhibitors of β-lactamases as an alternative approach to the control of resistant organisms has received considerable attention in recent years with the discovery of several powerful inhibitors. Methicillin, for example, is a fairly effective inhibitor of certain β-lactamases. By protecting lactamase-susceptible compounds from inactivation, methicillin gives a synergistic inhibitor effect when used in combination with other β-lactams although the effect is little used clinically. The discovery of the naturally occurring inhibitor, clavulanic acid (Fig. 2.16), has dramatically opened the way for clinically valuable synergism with β-lactamase-susceptible agents such as ampicillin and amoxycillin. Clavulanic acid, unlike methicillin, has little antibiotic activity of its own but it is a remarkably effective inhibitor of many of the β-lactamases of Gram-positive and Gram-negative bacteria. The chromosomally determined β-lactamases of certain Gram-negative bacteria which inactivate many cephalosporins are unfortunately not inhibited by clavulanic acid. Since clavulanic acid contains a β-lactam ring, the compound most probably attacks β-lactamase at or near its active centre to produce a covalently bound complex that is only slowly cleaved to release active enzyme. Clavulanic acid, therefore, behaves rather like a substrate with a low Michaelis constant but also with a very low V_{max}.

Origin of β-lactamases

There is ample evidence that penicillin- and cephalosporin-destroying enzymes existed in bacteria long before the development of these antibiotics by medical science. The widespread and often indiscriminate use of the penicillins has certainly encouraged an alarming rate of spread of resistant bacteria but at least clinical usage cannot be held responsible for the existence of the β-lactamases. If antibiotics, such as penicillin and cephalosporins, are released under natural conditions from the micro-organisms that produce them the synthesis of β-lactamases by other organisms sharing the same micro-environment might be regarded as the result of a long evolutionary process directed by the selected pressure of the antibiotics. In addition, organisms that produce β-lactams are equipped with β-lactamases, presumably as a self-defence measure.

It is possible that β-lactamases are the product of a mutant form of one of the genes coding for the penicillin/cephalosporin target enzymes, i.e. the transpeptidases involved in the final cross-linking stage of peptidoglycan biosynthesis and D-alanine carboxypeptidase (Chapter 2). β-Lactams form covalent complexes with these enzymes, but, although the complexes are usually sufficiently stable to ensure inhibition of peptidase function, there is evidence that in some cases the enzymes act as inefficient β-lactamases. True β-lactamases might have resulted from an evolutionary process that markedly enhanced the β-lactamase at the expense of the original transpeptidase activity. At some early stage there must have been duplication of a transpeptidase gene in order that the organism should retain the capacity for normal cell wall synthesis while evolving β-lactamase activity from mutations in the other gene copy.

CHLORAMPHENICOL

Another important example of bacterial drug resistance due to inactivation of the antibiotic is the enzymic acetylation of chloramphenicol by resistant strains of both Gram-positive and Gram-negative bacteria. The enzyme responsible, chloramphenicol acetyltransferase (CAT), is usually plasmid encoded and the gene for a major subtype is found on a transposon, Tn9 in Gram-negative bacteria. A chromosomally determined form occurs in *Pseudomonas mirabilis*. CATs are widely distributed in bacteria and despite differences in amino acid sequences, all CATs appear to be tetramers with identical subunits with molecular weights in the range 23 000 to 25 000. In Gram-positive bacteria the enzymes are inducible whereas in Gram-negative organisms enzyme synthesis is constitutive. As shown in Fig. 8.5 the enzyme first converts chloramphenicol to the 3-acetoxy derivative using acetyl-Co A as an essential cofactor. A non-enzymic intramolecular rearrangement then follows which transfers the acetoxy group to the 1-position. Further enzymic acetylation at the 3-position generates the final product, 1,3-diacetoxychloramphenicol. Both the mono- and di-acetoxy derivatives are inactive as antibiotics so that the two-step acetylation sequence is somewhat metabolically inefficient. Recently, several new derivatives of chlorampheni-

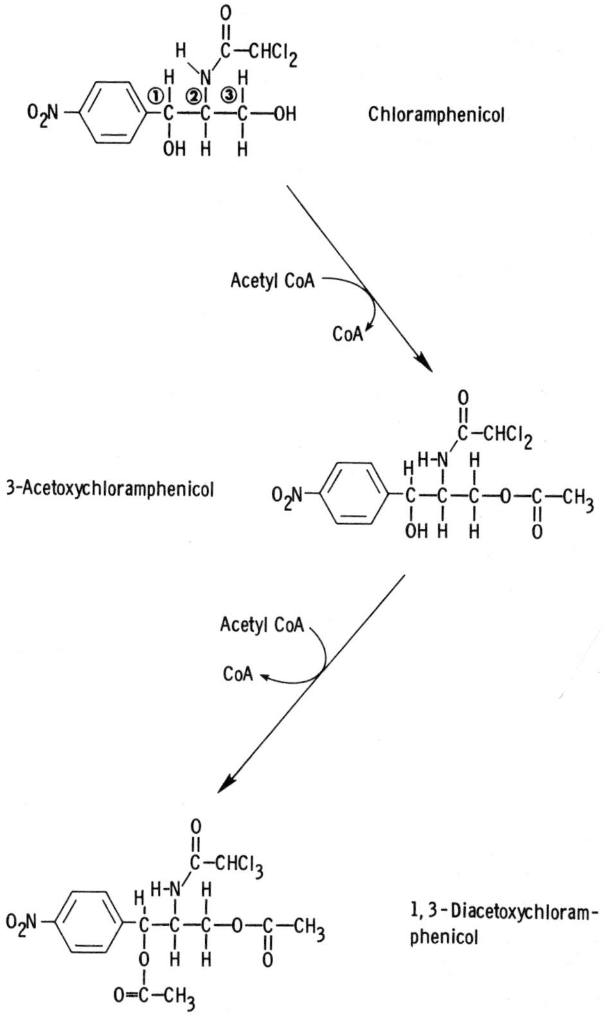

Figure 8.5 Inactivation of chloramphenicol by chloramphenicol acetyltransferase; 3-acetoxychloramphenicol is formed first and 1,3-diacetoxychloramphenicol probably forms by a non-enzymic shift of the acetyl group to the 1-position followed by acetylation of the free 3-OH group.

col have been described which are not subject to acetylation and yet retain antibacterial activity. The key substitution appears to be the replacement of the *terminal* OH group of the side chain with fluorine. None of these compounds has yet established a role in clinical medicine.

Physiology of chloramphenicol acetyltransferase synthesis

1. Gram-positive bacteria.   Kinetic studies of the induction process in *Staphylococcus aureus* show that in the early phase there is a conflict between the ability of chloramphenicol to induce the synthesis of CAT and its ability to inhibit protein biosynthesis. Eventually, sufficient enzyme is produced to depress the concentration of chloramphenicol below that needed to inhibit protein synthesis. Since this concentration of antibiotic still effectively induces enzyme synthesis, the production of chloramphenicol acetyltransferase now proceeds rapidly. An analogue of chloramphenicol, 3-deoxychloramphenicol, induces CAT but does not inhibit protein biosynthesis. Furthermore, it is not a substrate for chloramphenicol acetyltransferase. When 3-deoxychloramphenicol is used to induce CAT the biosynthesis of the enzyme proceeds without the prolonged lag that is observed with chloramphenicol.

The mechanism of CAT induction in *Staphylococcus aureus* is not clear. A simple repressor-controlled regulation seems unlikely since it has been impossible to isolate constitutive mutants.

2. Gram-negative bacteria.   Although CAT synthesis is constitutive in *Escherichia coli* it is subject to catabolite repression. Thus synthesis occurs at a higher rate in cultures grown on glycerol compared with glucose-supported cultures. There is also evidence to indicate that both cyclic AMP and catabolite activator protein are required for optimal synthesis of CAT. A complex of cyclic AMP and the activator protein is believed to interact with both the RNA polymerase and the promoter involved in CAT gene transcription.

AMINOGLYCOSIDES

The discovery of enzymes from aminoglycoside-resistant bacteria which convert aminoglycosides to biologically inert products seemed at first to place aminoglycoside resistance in the same category as resistance to β-lactams and chloramphenicol. More extensive investigations have revealed, however, that the basic principle of aminoglycoside resistance may be subtly different. Enzymic modification of aminoglycosides, which occurs in the periplasmic space or on the cytoplasmic membrane, has a double effect. Firstly the modified compound is no longer recognized by the facilitated transport process that permits rapid entry of drug into the cell (Chapter 7), and secondly the inhibitor action on cell growth and division is lost because the modified compound no longer affects the ribosomal target. The production of aminoglycoside-inactivating enzymes is low so that little of the extracellular drug is destroyed.

The aminoglycosides are a structurally diverse group of compounds. It is not surprising, therefore, that bacteria have evolved a considerable range of

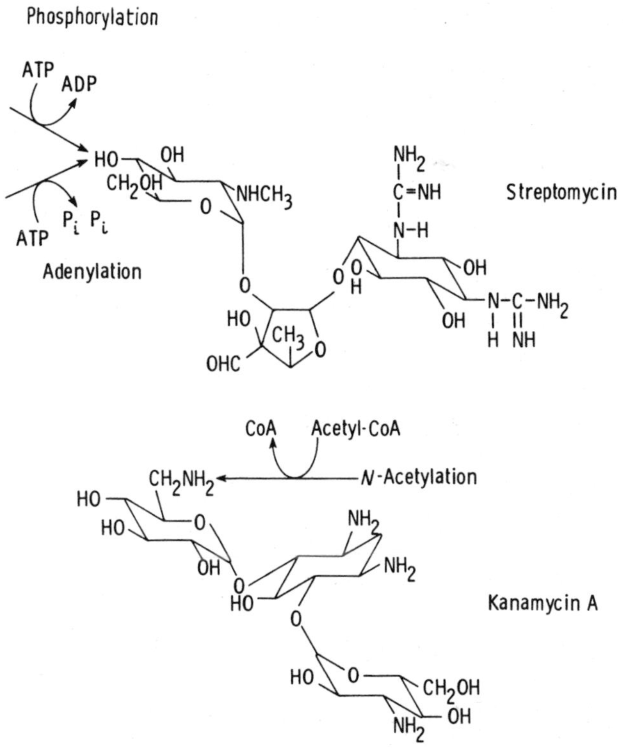

Phosphorylation

Figure 8.6 Three modes of enzymic inactivation of aminoglycosides. Streptomycin is not subject to N-acetylation, but kanamycin A is also inactivated by adenylation and phosphorylation.

defensive enzymes to deploy against them. Fortunately there are only three basic enzymic mechanisms of inactivation: (i) phosphorylation, (ii) adenylation and (iii) acetylation. Phosphorylation and adenylation occur at susceptible OH groups and acetylation at certain NH_2 groups. Enzymes carrying out these reactions have been detected in Gram-positive and Gram-negative bacteria. Some representative reactions are illustrated in Fig. 8.6.

In Gram-negative bacteria the genes for the aminoglycoside-modifying enzymes are plasmid-borne and several are found on transposons. Plasmids also carry the genes in Gram-positive organisms. Usually, the synthesis of aminoglycoside-modifying enzymes is constitutive, although this aspect of aminoglycoside resistance has not been studied in any detail. There is some evidence that synthesis of several of the enzymes is subject to catabolite repression.

For a comprehensive account of the various aminoglycoside-modifying enzymes the reader should consult a recent review mentioned at the end of this chapter. In summary, the three mechanisms involving phosphorylation, adenylation and acetylation can be further subdivided according to the specific OH and NH_2 groups that are involved in the reaction. Attempts have been made to subdivide the enzymes within the three mechanistic groups but these add little to our understanding of their diversity. Phosphorylation of susceptible OH groups requires ATP as the phosphate donor, and the same nucleotide provides AMP for the adenylation reaction. Streptomycin is a substrate for both an O-phosphotransferase and an O-nucleotidyltransferase, the same OH being attacked by each (Fig. 8.6). The widespread N-acetylation reaction requires the transfer of an acetyl group from acetyl-CoA to an NH_2 group and it only affects those antibiotics with a deoxystreptamine moiety (Fig. 8.6); streptomycin and spectinomycin are thus immune from this type of inactivation. N-Acetylation is not invariably associated with cellular resistance. Tobramycin, for example, is a relatively poor substrate for the N-acetyltransferase that readily inactivates gentamicin C_{1a} (the 3-amino group of the deoxystreptamine ring is attacked). Bacteria that produce this enzyme are resistant to gentamicin C_{1a} but are quite sensitive to tobramycin. Presumably tobramycin is not N-acetylated quickly enough to limit its entry into the bacterial cells. The emergence of another form of this N-acetyltransferase, which more effectively acetylates tobramycin, causes cellular resistance in pathogenic bacteria.

(2) Modification of the drug-sensitive site

STREPTOMYCIN

A striking example of the modification of a drug-sensitive site resulting in a high level of resistance is the loss of ribosomal sensitivity to streptomycin. The change in ribosomal structure has been traced to a single amino acid replacement in either of two positions in the S12 protein of the 30S ribosomal subunit. The changes in the S12 protein are caused by chromosomal mutations. The modified 30S subunit no longer binds streptomycin and the drug cannot therefore exert its characteristic effects on protein biosynthesis. The S12 protein may also be modified in mutants showing streptomycin dependence. Streptomycin resistance due to modified ribosomes is much less significant clinically than the plasmid-mediated resistance, although ribosomal resistance to streptomycin has been found in clinical isolates of *Neisseria gonorrhoeae*, *Staphylococcus aureus*, *Pseudomonas aeruginosa* and *Streptococcus faecalis*.

KASUGAMYCIN

This is an aminoglycoside antibiotic with an action on the 30S subunit of 70S ribosomes not unlike that of spectinomycin, i.e. it inhibits protein synthesis but it does not induce misreading nor can it cause phenotypic suppression. Kasugamycin-resistant mutants with resistant 30S subunits are unusual in having altered ribosomal RNA instead of protein. The mutation causes a loss in the activity of an enzyme which specifically methylates two adenine residues to dimethyladenine in the sequence, A-A-C-C-U-G, near the 3'-end of the 16S RNA. It is assumed that the change in the 16S RNA hinders the normal interaction between kasugamycin and the ribosome, but owing to the weak binding of the drug to sensitive ribosomes, it has not been possible to demonstrate unequivocally that there is diminished binding to resistant ribosomes. It is interesting that 30S subunits from kasugamycin-resistant cells are methylated by the methylase from sensitive cells. When these methylated subunits are recombined with 50S subunits the resultant 70S particles are sensitive to kasugamycin.

ERYTHROMYCIN

In natural isolates of *Staphylococcus aureus* and *Streptococcus* spp. resistance to erythromycin is associated with a change in the 23S RNA of the 50S ribosomal subunit. A plasmid-mediated ribosomal RNA methylase induced by erythromycin is responsible for the alteration in the 23S RNA which involves N-dimethylation of a specific adenine moiety using S-adenosylmethionine as the methyl donor. The chemical modification of the adenine residue is believed to induce a conformational change in the ribosome that diminishes its affinity for erythromycin. The modified ribosomes are cross-resistant to lincomycin. This mechanism of resistance provides an intriguing contrast with that of kasugamycin where a specific *loss* of RNA methylation is involved.

The inductive mechanism does not involve the interaction of erythromycin with a repressor protein and a subsequent increase in transcription but rather an increase in the rate of translation of the mRNA for the N-dimethylase. In uninduced cells translation of the message is hindered by specific secondary structure in the mRNA which is indirectly relieved by the action of erythromycin on the ribosomes. Paradoxically this permits limited biosynthesis of the N-dimethylase. As the methylated ribosomes escape the inhibitor effects of erythromycin, translation of the message for N-dimethylase slows once again.

RIFAMYCINS

Another important example of a mutation causing a change in the drug-sensitive site is seen in the altered DNA-dependent RNA polymerase of mutants resistant to the rifamycins. As we saw in Chapter 4, resistance to the rifamycins depends on a modification of the β-subunit of the 'core' enzyme of the polymerase. Loss of sensitivity of the polymerase to rifampicin is accompanied by a failure of the 'core' enzyme to bind the antibiotic. The change in the enzyme is determined by a chromosomal mutation.

QUINOLONES

Chromosomal mutations affecting the A subunit of the target enzyme DNA gyrase confer resistance in *Escherichia coli* to nalidixic acid and oxolinic acid. Mutations in the B subunit cause resistance to novobiocin and coumermycin. Interestingly, nalidixic acid-resistant mutants remain sensitive to another quinolone, ciprofloxacin, probably because of the very high potency of this compound.

It is not possible for an organism to acquire resistance to every inhibitor by modification of the appropriate target site. Some changes that would confer resistance are almost certainly incompatible with cell viability. The consistent failure to isolate bacterial mutants with tetracycline-resistant ribosomes, for example, could be explained if such resistant ribosomes were non-functional in protein biosynthesis.

(3) Loss of cell permeability to a drug

Several mechanisms of resistance mimic a fall in cell permeability to a drug. A familiar example occurs where resistant cells convert a drug to a derivative that does not bind to its intracellular target. The cells no longer sequester the drug and a false impression of diminished permeability is created. Similarly, a change in the target site leading to a loss of affinity for a drug results in a reduced cellular uptake of drug.

Despite these traps for the unwary investigator, resistance to certain antimicrobial drugs clearly depends on cellular changes that hinder drug access. This loss of permeability may be due to one of several mechanisms. (i) An existing permeability barrier may be modified. The greater intrinsic antibiotic insensitivity of Gram-negative bacteria compared with Gram-positive organisms depends on the non-specific permeability barrier provided by the outer membrane of Gram-negative cells (Chapter 7). Mutations affecting the lipopolysaccharides, porins and nutrient receptor proteins of the outer membrane may all adversely affect the inward movement of antibiotics. These changes may not in themselves always give rise to significant resistance, but combined with other factors, such as drug-inactivating enzymes or a reduction in target sensitivity, they may be sufficient to afford the cells adequate protection. The reduced uptake of streptomycin and erythromycin by some resistant pneumococci may be due to the development of an altered capsular permeability barrier. Since these strains are less susceptible to DNA transformation because of a diminished uptake of DNA, the permeability barrier is probably non-specific.

(ii) When the drug gains intracellular access by a specific transport mechanism, diminished uptake by resistant cells can be due to a mutation that results in a partial or complete loss of transport function. As we saw in Chapter 7, drugs may subvert permeation mechanisms designed for nutrients. A mutation affecting such a permeation mechanism may give rise to resistance if the cells can withstand or circumvent the loss of the nutrient. A good example is found in a series of mutants of *Escherichia coli* resistant to D-cycloserine. In sensitive cells this drug is accumulated by the amino acid-transport system that normally carries D-alanine or glycine. Mutants exhibiting graded loss of function of this transport system are resistant to cycloserine. Resistance to the clinically valuable antifungal drug 5-fluorocytosine (Chapter 6) can arise from mutations affecting the activity of cytosine permease. This enzyme is exploited by 5-fluorocytosine in order to penetrate the fungal cell membrane.

(iii) Specific antagonism of antibiotic transport. The most commonly encountered mode of resistance to tetracyclines in Gram-positive and Gram-negative bacteria depends on a diminished cellular accumulation of these compounds. Many tetracycline-sensitive bacteria accumulate tetracyclines by an energy-dependent process which involves the cytoplasmic membrane (Chapter 7). Tetracycline accumulation in resistant cells is blocked by a partially inducible system that is determined by plasmid genes in Gram-positive and Gram-negative bacteria. When resistant cells are exposed to a subinhibitory concentration of tetracycline there is a rapid increase in the level of resistance which can be prevented by blocking protein or RNA synthesis in the cells. The increase in resistance is associated with a fall in the uptake of tetracycline by the cells. In enteric bacteria two transposons, Tn1721 and Tn10, commonly carry the genes for tetracycline resistance. Tn10 encodes two proteins: a membrane-located protein of molecular weight 36 000 that appears to mediate resistance and a repressor protein associated with the inducibility of tetracycline resistance. In Gram-positive bacteria the induction mechanism does not appear to involve a repressor protein but depends upon a translational attenuation mechanism of the type previously described for erythromycin resistance.

The actual basis for tetracycline resistance has proved to be an efflux process that specifically pumps tetracycline out of the cells against a concentration gradient. The pump, presumably associated with the inducible membrane protein, is therefore energy-dependent and appears to harness the proton motive force across the cell membrane. Although the mechanism of tetracycline resistance is essentially the same in Gram-positive and Gram-negative bacteria, there is some evidence that the proteins concerned may have evolved separately in the two genera.

DRUG EFFLUX IN TUMOUR CELLS

An interesting parallel with the tetracycline efflux system has recently emerged from studies on drug-resistant tumour cells. Treatment failures with anticancer drugs are often due to the selection of drug-resistant cells. Treatment with a single drug sometimes results in the development of a tumour cell population resistant not only to the treatment drug but also to a range of chemically unrelated anticancer agents. This intriguing but potentially devastating phenomenon has been traced to a membrane glycoprotein that removes cytotoxic compounds from the cytoplasm by an energy-dependent efflux mechanism. The gene coding for this protein, which occurs in some normal cells, is greatly amplified in resistant tumour cells leading to a greatly enhanced production of the drug-effluxing protein. The protein comprises two homologous halves, each half containing hydrophobic and hydrophilic regions. Quite remarkably, the hydrophilic region shows extensive sequence homology with the ATP-binding component of the bacterial nutrient transport proteins associated with the cytoplasmic membrane.

(4) Synthesis of an additional drug-resistant enzyme or overproduction of drug-sensitive enzyme

SULPHONAMIDES

Resistance to the sulphonamides has been recognized for nearly forty years and the resistance determinant was found on R-plasmids when they were discovered in Japan in the 1950s. The plasmids carry the genetic information for two forms of dihydropteroate synthase which are highly resistant to inhibition by sulphonamides. The chromosomally determined enzyme (Chapter 6) is the target of the sulphonamides. R-plasmid-determined dihydropteroate synthase binds sulphonamides 10 000 times less tightly than the chromosomal enzyme, while the K_m for the substrate, p-aminobenzoic acid, is the same for both enzymes. Cells with the additional resistant enzyme utilize p-aminobenzoic acid normally, despite high concentrations of sulphonamides. Pneumococci resistant to sulphonamides also synthesize a dihydropteroate synthase that is not inhibited by sulphonamides.

TRIMETHOPRIM

The sulphonamide, sulphamethoxazole, is often used in combination with trimethoprim (Chapter 6) because it was believed that resistance to this combination would be slow to emerge. However, bacteria causing serious infections are now appearing equipped with plasmid-borne genes for both sulphonamide and trimethoprim resistance. The mechanism of trimethoprim resistance follows the same principle as sulphonamide resistance; the plasmid carries a gene which directs the synthesis of a dihydrofolate reductase that is highly resistant to trimethoprim. The K_m for the substrate, dihydrofolate, is similar to that of the chromosomally determined trimethoprim-sensitive enzyme. In fact there are four forms of trimethoprim-resistant enzymes determined by R-plasmids. Type I enzymes are several thousand times more resistant to trimethoprim than the chromosomal enzyme, whilst Type II enzymes are almost totally resistant to trimethoprim. Interestingly, both types of enzyme are highly resistant to methotrexate which is a close structural analogue of dihydrofolate. These enzymes mediate high level resistance to trimethoprim. In contrast, the Type III enzyme provides only a moderate increase in resistance, since it is inhibited by relatively low concentrations of trimethoprim. The concentration required to achieve 50% inhibition of the Type III enzyme is $1.5\,\mu$M compared with $0.007\,\mu$M for the chromosomal enzyme. A fourth type of plasmid-determined dihydrofolate reductase differs markedly from the other enzymes. Its susceptibility to trimethoprim is only slightly less than that of the chromosomal enzyme but its rate of synthesis is dramatically increased by the inhibitor. The induced rate of synthesis is up to 600 times greater than the resting level and almost certainly underlies the enhanced resistance to trimethoprim of bacteria that carry this plasmid.

METHICILLIN

Another example of drug-resistance dependent upon the production of a modified form of the target enzyme is found in the potentially dangerous methicillin-resistant strains of *Staphylococcus aureus* that plague many hospitals. Most of these strains have the ability to make a unique penicillin-binding protein, PBP2a, which has a lower affinity for methicillin and β-lactams in general than the usual form of this protein. The production of PBP2a appears to be stimulated by the presence of methicillin although the regulatory mechanism involved is unknown. The gene encoding PBP2a may be located on a transposon.

(5) Increased production of a metabolite that antagonizes the inhibitor

When a drug inhibits growth by the competitive antagonism of a normal metabolite, resistance to such an inhibitor may be due to increased production of the metabolite; the inhibitor is competitively displaced from its enzyme binding site. This type of resistance is exemplified by certain mutants resistant to sulphonamides. In these cells the concentration of *p*-aminobenzoic acid is said to be substantially higher than in sulphonamide-sensitive cells although the mechanism underlying the increased production of *p*-aminobenzoic acid is uncertain. Sulphonamide resistance due to resistant dihydropteroate synthase is probably much more important in the clinical setting.

BACTERIAL TOLERANCE

A special type of resistance to β-lactam antibiotics has been encountered amongst clinical isolates in recent years. The β-lactams are characterized by their very useful ability to kill growing bacteria. However, below the concentrations required for bactericidal activity β-lactams also arrest cell growth. β-Lactam-tolerant mutants show no change in their sensitivity to the growth-inhibiting action of β-lactams but are much more resistant to their bactericidal activity than the parent organisms. The consequence for the patient is that the course of their infection and treatment period is unduly prolonged since the successful application of β-lactam therapy depends to a considerable extent upon the expected bactericidal component. The basis for β-lactam tolerance does not appear to rest upon β-lactamase activity or changes in the sensitivity of the molecular targets of β-lactams but stems from a decline in the normal drug-induced disruption of the structure of the expanding peptidoglycan envelope of growing bacteria. This disruption leads to the inappropriate activation of autolysins, or peptidoglycan hydrolases which cause disintegration of the bacterial cell wall and rapid cell death. β-Lactam-tolerant mutants have a defect in their autolysin activity which, though causing some abnormality in cell morphology, affords protection against the lethal effects of β-lactams. Such tolerant organisms have been reported amongst clinical isolates of *Pneumococci* and other Gram-positive bacteria. There is little evidence so far to suggest that the β-lactam tolerance poses a major threat to the successful treatment of Gram-negative infections.

PRACTICAL APPROACHES TO THE CONTROL OF DRUG RESISTANCE

We have reviewed the genetic and biochemical ingenuity of micro-organisms in combating the toxic effects of growth-inhibiting substances. Highly successful mechanisms of resistance against antibiotics were evolved by micro-organisms long before the introduction of these substances into medical and veterinary practice. The development of resistance against novel chemically synthesized agents unrelated to antibiotics, natural metabolites or growth factors is more difficult to explain, although the high rate of cell division in most microbial populations combined with their extraordinary genetic flexibility facilitates 'high speed' evolution. Nevertheless, the capacity of bacteria to develop resistance is not unlimited. Clinical experience with β-lactamase-stable derivatives gives some support to this view, since resistance to these agents has appeared much more slowly than to the original penicillin. It is also interesting that streptococci apparently do not become resistant to

β-lactam antibiotics under practical conditions of use. A detailed consideration of the common mechanisms of resistance and their frequency of occurrence among important pathogens can assist the design of new chemotherapeutic agents. Considerable efforts continue to be directed to the development of new β-lactams with improved stability to the many different
bacterial β-lactamases. The increasing emphasis on the chemotherapy of
fungal and viral infections must raise concerns about the likelihood of
treatment failures arising from the emergence of drug-resistant fungi and
viruses.

Apart from the preparation of new chemical agents there are a number of
other measures which can limit the problem of drug resistance.

(1) Treatment of an infection usually begins with the administration of a
high 'loading' dose of antimicrobial drug in the hope of eliminating resistant
organisms in the body. Therapy is continued with doses large enough and
frequent enough to maintain an effective antibacterial level of drug in the
blood until the infection is overcome.

(2) Micro-organisms are often cross-resistant to chemically related drugs.
Simultaneous therapy with chemically unrelated inhibitors, however, presents a more formidable challenge to a micro-organism especially when the
inhibitors attack essential biochemistry at two different points. Although
R-plasmids conferring resistance to the sulphamethoxazole–trimethoprim
combination are no longer uncommon, several years of 'grace' were obtained
before this occurred. In the treatment of tuberculosis, where the need for
therapy extending over many months encourages the development of drug
resistance, the use of a combination of isoniazid, streptomycin and p-
aminosalicylic acid successfully avoids the frequent emergence of resistant
strains. The clinical use of combinations of inhibitors of β-lactamase and
β-lactams is proving to be useful in countering the threat of β-lactamase-
producing pathogens.

(3) Undoubtedly, the careless use of antimicrobial agents in human
medicine and in farming for non-therapeutic purposes has assisted the
spread of resistant organisms. An encouraging development, however, is
towards much stricter control over the use of these drugs. The proper and
careful deployment of antimicrobial agents can produce dramatic results in
reducing the number of resistant organisms in the environment.

FURTHER READING

Bryan, L. E. (ed.) (1984) *Antimicrobial Drug Resistance*, Academic Press, London.
Clewell, D. B. and Gawron-Burke, C. (1986) Conjugative transposons and dissemi-
nation of antibiotic resistance in *Streptococci. Annu. Rev. Microbiol.*, **40**, 635.
Datta, N. (ed.) (1984) Antibiotic resistance in bacteria. *Brit. Med. Bull.*, **40**, 1.
Foster, T. J. (1983) Plasmid-determined resistance to antimicrobial drugs and toxic
metal ions in bacteria. *Microbiol. Rev.*, **47**, 361.

Handwerger, S. and Tomasz, A. (1985) Antibiotic tolerance among clinical isolates of bacteria. *Annu. Rev. Pharmacol. Toxicol.*, **25**, 349.

Iwata, K. (1986) Drug resistance in human pathogenic fungi. In *In Vitro and In Vivo Evaluation of Antifungal Agents* (eds K. Iwata and H. van den Bossche), Elsevier Science Publishers, Amsterdam.

Jacoby, G. A. (1985) Genetics and epidemiology of resistance. In *The Scientific Basis of Antimicrobial Therapy, Society for General Microbiology Symposium 38* (eds D. Greenwood and F. O'Grady).

Lyon, B. R. and Skurray, R. (1987) Antimicrobial resistance of *Staphylococcus aureus*: genetic basis. *Microbiol. Rev.*, **51**, 85.

Sanders, C. C. (1987) Chromosomal cephalosporins responsible for multiple resistance to newer β-lactam antibiotics. *Annu. Rev. Microbiol.*, **41**, 573.

Watson, J. D., Hopkins, N. H., Roberts, J. W., Steitz, J. A. and Weiner, A. M. (1987) *Molecular Biology of the Gene*, Benjamin/Cummings Publishing Co.

Wiedemann, B., Bennett, P. M., Linton, A. H., Sköld, O. and Speller, D. C. E. (eds) (1986) Evolution, ecology and epidemiology of antibiotic resistance. *J. Antimicrob. Ther.*, **18**, (Supplement C).

Willetts, N. and Wilkins, B. (1984) Processing of plasmid DNA during bacterial conjugation. *Microbiol. Rev.*, **48**, 24.

Index

Arabinosyl adenine
 inhibition of DNA synthesis 157
 structure 157
 virus inhibitor 157
Arildone
 inhibition of uncoating of viruses 156
 structure 155
Arsenicals
 binding to lipoic acid 138
 inhibition of oxidative decarboxylation
  of α-keto acids 138
 separation of toxicity and
  chemotherapeutic activity 6
Arsenious oxide, as trypanocide 5
Artemisinie, *see* Qinghaosu
Asepsis 4
Asulam
 herbicide 77
 structure 75
Atebrin, *see* Mepacrine
Atoxyl
 resistance in trypanosomes 8
 structure 5
Auxotrophic bacteria, use in determining
  site of antibacterial action 15
Azaserine
 clinical use 81
 inhibition of purine nucleotide
  synthesis 81, 82
 radiomimetic and mutagenic
  properties 82
 structure 81
Azidothymidine
 biochemical action 157, 158
 structure 157
 use in AIDS 157
Azole antifungals
 accumulation of methylsterols 145
 clinical use 145
 demethylation 145
 effect on chitin synthesis 145
 inhibition of sterol C14 145
 interaction with cytochrome P_{450} 145
 structural types 147
AZT, *see* Azidothymidine

BW58C80
 antiparasitic action 139
 structure 138
Bacitracin, binding to pyrophosphates
  37
 complex formation with farnesyl
  pyrophosphate 37
 complex formation with undecaprenyl
  pyrophosphate 37
 structure 37
 uses and antibacterial action 36

Bacterial cell wall
 permeability function 161, 164, 165,
  166, 167
 protective function 20
Bacterial cell, differences from animal
  cells 20
Bacterial tolerance 198
Bacteriostatic and bactericidal action
  compared 55, 56
Bacteriostatic drugs in the treatment of
  systemic infections 55
Benomyl
 agricultural fungicide 149
 effect on chromosome separation
  149
 metabolism to MBC 149
 structure 149
Benzyl penicillin
 affinity for β-lactamase 186
Biocides 55
Biosynthetic pathways, methods of
  determining sites of blockage 15
Bleomycin A_1 structure 97
Bleomycins, action on DNA 96
Bone marrow, toxic effects of
  chloramphenicol on 135
Borrelidin 121
Bridging groups, variation in different
  types of peptidoglycan 33
Broad spectrum antibiotics, definition
  120, 121
Bronopol
 effects on membranes 58

Calcium
 binding to A23187 69
 significance in Gram-negative
  bacterial wall 165
Carbobenzoxyleucylchloromethylketone
 inhibition of virus protease 158
 structure 158
Carboxin
 agricultural fungicide 153
 blocks succinate branch of respiratory
  pathway 153
 structure 153
Carboxypeptidases I and II 34
Carboxypeptidases
 active centre 51
 extracellular, from streptomycetes 50
 reaction intermediates 51
 with transpeptidase activity 50
Castanospermine
 inhibition of AIDS virus 159
 structure 158
Cationic antiseptics 59, 60
Cefoxitin 42